Energy, Climate and the Environment Series
Series Editor: David Elliott, Emeritus Professor of Technology,
Open University, UK

Titles include:

Luca Anceschi and Jonathan Symons (*editors*)
ENERGY SECURITY IN THE ERA OF CLIMATE CHANGE
The Asia-Pacific Experience

Ian Bailey and Hugh Compston (*editors*)
FEELING THE HEAT
The Politics of Climate Policy in Rapidly Industrializing Countries

Mehmet Efe Biresselioglu
EUROPEAN ENERGY SECURITY
Turkey's Future Role and Impact

David Elliott (*editor*)
NUCLEAR OR NOT?
Does Nuclear Power Have a Place in a Sustainable Future?

David Elliott (*editor*)
SUSTAINABLE ENERGY
Opportunities and Limitations

Horace Herring and Steve Sorrell (*editors*)
ENERGY EFFICIENCY AND SUSTAINABLE CONSUMPTION
The Rebound Effect

Horace Herring (*editor*)
LIVING IN A LOW-CARBON SOCIETY IN 2050

Matti Kojo and Tapio Litmanen (*editors*)
THE RENEWAL OF NUCLEAR POWER IN FINLAND

Antonio Marquina (*editor*)
GLOBAL WARMING AND CLIMATE CHANGE
Prospects and Policies in Asia and Europe

Catherine Mitchell
THE POLITICAL ECONOMY OF SUSTAINABLE ENERGY

Ivan Scrase and Gordon MacKerron (*editors*)
ENERGY FOR THE FUTURE
A New Agenda

Gill Seyfang
SUSTAINABLE CONSUMPTION, COMMUNITY ACTION
AND THE NEW ECONOMICS
Seeds of Change

Joseph Szarka
WIND POWER IN EUROPE
Politics, Business and Society

Joseph Szarka, Richard Cowell, Geraint Ellis, Peter Strachan
and Charles Warren (editors)
LEARNING FROM WIND POWER
Governance, Societal and Policy Perspectives on Sustainable Energy

David Toke
ECOLOGICAL MODERNISATION AND RENEWABLE ENERGY

Xu Yi-chong (editor)
NUCLEAR ENERGY DEVELOPMENT IN ASIA
Problems and Prospects

Xu Yi-chong
THE POLITICS OF NUCLEAR ENERGY IN CHINA

Energy, Climate and the Environment
Series Standing Order ISBN 978–0–230–00800–7 (hardback)
978–0–230–22150–5 (paperback)

You can receive future titles in this series as they are published by placing a
standing order. Please contact your bookseller or, in case of difficulty, write to us
at the address below with your name and address, the title of the series and the
ISBNs quoted above.

Customer Services Department, Macmillan Distribution Ltd, Houndmills,
Basingstoke, Hampshire RG21 6XS, England

Living in a Low-Carbon Society in 2050

Edited by

Horace Herring
Visiting Research Fellow, Energy & Environment Research Unit,
The Open University, UK

palgrave
macmillan

First published 2012 by
PALGRAVE MACMILLAN

Palgrave Macmillan in the UK is an imprint of Macmillan Publishers Limited,
registered in England, company number 785998, of Houndmills, Basingstoke,
Hampshire RG21 6XS.

Palgrave Macmillan in the US is a division of St Martin's Press LLC,
175 Fifth Avenue, New York, NY 10010.

Palgrave Macmillan is the global academic imprint of the above companies
and has companies and representatives throughout the world.

Palgrave® and Macmillan® are registered trademarks in the United States,
the United Kingdom, Europe and other countries.

ISBN 978–0–230–28225–4

A catalogue record for this book is available from the British Library.

A catalog record for this book is available from the Library of Congress.

10   9   8   7   6   5   4   3   2   1
21   20   19   18   17   16   15   14   13   12

Transferred to Digital Printing in 2013

Contents

List of Tables and Figures		vii
Series Editor Preface		ix
Notes on Contributors		xii

1 Introduction   *Horace Herring* — 1

Part I Policy

2 What is a Low-Carbon Society?   *Stephen Peake* — 15

3 Low-Carbon Living in 2050   *Nicola Hole* — 28

4 What is the Carbon Footprint of a Decent Life?   *Angela Druckman and Tim Jackson* — 41

5 Transport and Mobility Choices in 2050   *Stephen Potter* — 60

6 A Low-Carbon Transition   *Neil Strachan and Timothy J. Foxon* — 75

Part II Case Studies

7 Case Studies in Low-Carbon Living   *Robin Roy* — 95

8 Designing and Creating my Low-Carbon Home   *Catherine Mitchell* — 121

9 Land Use Scenario 2050   *Grace Crabb and Adam Thorogood* — 129

10 Zero-Carbon Britain: Time to say 'we will'   *Tanya Hawkes* — 149

11 Low-Carbon Society in Switzerland   *Bastien Girod* — 164

Part III   Stories

12 Little Greenham   183
    James Goodman

13 The Housing Ladder   190
    Roger Levett

14 The Refugee   198
    Tanya Hawkes

15 The Gun and the Sun   210
    Horace Herring

16 Conclusion   218
    Horace Herring

Index   229

List of Tables and Figures

Tables

4.1   Housing specifications and heating regimes   44

4.2   Distribution of minimum acceptable standard
      of living budget family types   47

5.1   UK CO_2 emissions by source   61

5.2   Indices of transport trends   66

5.3   Current energy use of transport modes   69

7.1   The low-carbon living cases compared   113

9.1   Annual quantity of product range from the South-East
      lowland farm   138

9.2   Land practice carbon flux at South-East lowland farm   139

9.3   Annual quantity of product range from the
      Cambrian farm   141

9.4   Land practice carbon flux at the Cambrian farm   142

16.1 London-Cyprus return by fast and slow travel   224

Figures

2.1   The direction of evolution of carbon and energy in society   16

2.2   Summary of 22 'test-bed' low-carbon community
      projects in the UK   23

2.3   Growthist, peakist and environmentalist
      perspectives on low-carbon societies   24

3.1   A multi-level perspective for addressing the factors
      that affect energy consumption   32

4.1   Comparison of JRF expenditure budgets
      against expenditure   46

4.2   Comparison of mean GHG emissions per
      household for UK mean in 2004 against the
      minimum income standard emissions   48

4.3 Minimum income standard GHG emissions
compared against UK GHG reductions targets 49

6.1 Possible transition pathways and the factors that
influence them 83

6.2 A co-evolving model of the relationship between policy,
behaviour change and technology 85

7.1 Breakdown of all the direct and indirect carbon
dioxide emissions 96

7.2 (a–c) Oxford eco-renovated house 99

7.3 Annual gas and electricity use in the Yellow House
have halved since eco-renovation 101

7.4 (a–d) Front of the Yellow House 102

7.5 (a–b) One of the larger houses at Millennium Green 105

7.6 (a–c) Cross-section of the Autonomous House 106

7.7 (a–d) The Hockerton Housing project 109

7.8 (a–c) BedZED, South London 111

9.1 Hectares required for each system 137

11.1 Swiss greenhouse gas and CO_2 emissions for
different sectors 165

11.2 The Forum Chriesbach, a five-storey office and
research building for EAWAG 167

11.3 CO_2 emissions of different transportation modes 169

Series Editor Preface

Concerns about the potential environmental, social and economic impacts of climate change have led to a major international debate over what could and should be done to reduce emissions of greenhouse gases, which are claimed to be the main cause. There is still a scientific debate over the likely scale of climate change, and the complex interactions between human activities and climate systems, but, in the words of no less than the (then) Governor of California, Arnold Schwarzenegger, 'I say the debate is over. We know the science, we see the threat, and the time for action is now.'

Whatever we now do, there will have to be a lot of social and economic adaptation to climate change – preparing for increased flooding and other climate-related problems. However, the more fundamental response is to try to reduce or avoid the human activities that are seen as causing climate change. That means, primarily, trying to reduce or eliminate emission of greenhouse gases from the combustion of fossil fuels in vehicles, houses and power stations. Given that around 80 per cent of the energy used in the world at present comes from these sources, this will be a major technological, economic and political undertaking. It will involve reducing demand for energy (via lifestyle choice changes), producing and using whatever energy we still need more efficiently (getting more from less), and supplying the reduced amount of energy from non-fossil sources (basically switching over to renewables and/or nuclear power).

Each of these options opens up a range of social, economic and environmental issues. Industrial society and modern consumer cultures have been based on the ever-expanding use of fossil fuels, so the changes required will inevitably be challenging. Perhaps equally inevitable are disagreements and conflicts over the merits and demerits of the various options and in relation to strategies and policies for pursuing them. These conflicts and associated debates sometimes concern technical issues, but there are usually also underlying political and ideological commitments and agendas, which shape, or at least colour, the ostensibly technical debates. In particular, at times, technical assertions can be used to buttress specific policy frameworks in ways which subsequently prove to be flawed.

The aim of this series is to provide texts which lay out the technical, environmental and political issues relating to the various proposed policies for responding to climate change. The focus is not primarily on the science of climate change, or on the technological detail, although there will be accounts of the state of the art, to aid assessment of the viability of the various options. However, the main focus is the policy conflicts over which strategy to pursue. The series adopts a critical approach and attempts to identify flaws in emerging policies, propositions and assertions. In particular, it seeks to illuminate counter-intuitive assessments, conclusions and new perspectives. The aim is not simply to map the debates but to explore their structure, their underlying assumptions and their limitations. Texts are incisive and authoritative sources of critical analysis and commentary, indicating clearly the divergent views that have emerged and also identifying the shortcomings of these views. However, the books do not simply provide an overview, they also offer policy prescriptions.

The present volume tackles perhaps the most complex and certainly one of the most contentious issues – how can we change our lifestyles to reduce carbon emissions? It takes for granted that whatever we can do on the technology side, we also have to change how we live, but accepts that there are major debates about how much change is needed and about how it might be achieved.

To move things forward, this book looks at what a low-carbon society might look like, approaching this partly through traditional analysis and cases studies, but also, innovatively, through a series of short fictional stories, to try to catch some of the subjective reality and the human qualities of what life might be like in the future. Some of the stories are inspiring and optimistic, but some of the scenarios seem bleak – assuming that technology cannot really help much, so that drastic carbon constraints are imposed.

Is that inevitable? Over the last couple of years scenarios have emerged which, assuming significant energy conservation measures, have renewables supplying 95–100 per cent of all electricity and most energy by 2050 (for the EU and maybe globally), earlier in some cases. Obviously this may not happen, and certainly we ought to try to cut demand as much as possible in any case – that makes meeting it from renewables easier, and often the cheapest option.

This book assumes some serious energy saving through technical upgrades, especially to houses, as well as a lot of new large-scale green energy generation, transmission and storage infrastructure (for grid balancing), as well as local level initiatives. But its main focus is on what can be done by individuals and in houses for energy, as well as in terms of transport and food supply. Within that mainly domestic context, it is pretty sobering: lifestyles and expectations may have to change radically.

While there is the risk of pandering to the view that we can only deal with climate change if we adopt a frugal lifestyle, the opposite 'renewables as technical fix' view is also risky – if demand is to be constrained, lifestyle changes are needed, in the affluent countries in particular. This book gives a good sense of what changes might be required and how they might feel in reality, but the questions however remain, how much change is needed and how soon?

Notes on Contributors

Grace Crabb has worked in practical water and land management for ten years, both in the UK and abroad. For the last five years she has been teaching and doing research on ecological sanitation and sustainable land management at the Centre for Alternative Technology in Wales.

Angela Druckman is Senior Lecturer in Sustainable Energy and Climate Change Mitigation at the University of Surrey, and is a member of the ESRC Research group on Lifestyles, Values and Environment (RESOLVE).

Timothy Foxon is an RCUK Academic Fellow in the Sustainability Research Institute at the University of Leeds, and a member of the ESRC Centre for Climate Change, Economics and Policy.

Bastien Girod holds a Master's degree in environmental sciences from the Swiss Federal Institute of Technology (ETH Zurich), Switzerland. At the same institution he did his PhD on the integration of rebound effect into LCA. He is currently pursuing his Post Doc at the Utrecht Center for Earth and Sustainability of the University Utrecht. His research focus includes GHG emissions from households, long-term emission scenarios and the challenge of societal transition towards sustainable development. Since 2007, he is also member of the Swiss National Parliament and Swiss commission for environmental legislation.

James Goodman is Head of Futures with the independent sustainability experts Forum for the Future (www.forumforethefuture.org).

Tanya Hawkes lectures at the Centre for Alternative Technology, Machynlleth, Wales, on national and international climate change policy and works at Platform and the Public Interest Research Centre.

Horace Herring is a freelance writer, editor and researcher, specialising in energy efficiency, and sustainable consumption and communities. He has been 'visiting' the Open University since 1993.

Nicola Hole is a doctoral student in the energy policy group at the University of Exeter. She is currently investigating the social complexities of modern lifestyles that lead to specific energy behaviours and exploring the comprehensive changes to lifestyles and practices that are needed to reduce energy demand.

Tim Jackson is Professor of Sustainable Development at the University of Surrey, Director of the Sustainable Lifestyles Research Group and author of the controversial and groundbreaking book *Prosperity Without Growth*. He is also an award-winning playwright with numerous radio-writing credits for the BBC.

Roger Levett is a partner in Levett-Therivel sustainability consultants and specialises in public policy, planning and appraisal for sustainability.

Catherine Mitchell is Professor of Energy Policy at the University of Exeter and has worked extensively on energy policy development as an academic, within government and as an advisor to various national and international institutions, and companies. She is currently a Lead Author in the IPCC's 5th Assessment Report.

Stephen Peake is Senior Lecturer in Environmental Technology at the Open University and Fellow of the Judge Business School, University of Cambridge.

Stephen Potter is Professor of Transport Strategy at the Open University, and researches the design processes involved for the diffusion of cleaner transport and vehicle technologies, and more sustainable travel behaviour. He also undertakes studies in other sustainable design issues, including low-carbon domestic technologies and energy systems.

Robin Roy is Professor of Design and Environment at the Open University. In 1979 he founded the Design Innovation Group to conduct research on product development, innovation and sustainable design, and has contributed to many OU distance-teaching courses on design, technology and the environment.

Neil Strachan is an interdisciplinary energy economist and a reader (Associate Professor) in Energy Economics and Modelling at University College London.

Adam Thorogood works for the Centre for Alternative Technology, and co-directs an environmental social enterprise, Dyfi Woodlands, and is a director of Wales' community woodland association, Llais y Goedwig. He is interested in the impacts of climate change on forest ecosystems.

1
Introduction

Horace Herring

This is a book about living in a low-carbon society. There are already hundreds of millions of people leading such a lifestyle, but this book is not aimed at them. These people are, of course, the world's poor, living on a few dollars a day and unable to afford fossil fuels, and they subsist mainly on biomass. Their fervent desire is to join our current carbon-intensive society and enjoy its wealth of consumption.

Rather, this book is aimed at those who want to reduce their carbon emissions but still keep an affluent lifestyle; for those who want to be low-carbon but still rich. But is this possible? Technically, of course, yes, but is it economically feasible and socially desirable?

Will it involve too much of a lifestyle change, of having to give up stuff, of going without? Governments may talk about reducing carbon dioxide (CO_2) emissions by 80 per cent by 2050, but 20 years of carbon-reduction efforts, since 1990, have achieved little. True, we have achieved the European target of an 8 per cent decrease in greenhouse gas emissions, but nearly half that cut has been due to big reductions in the minor greenhouse gases, such as methane and nitrous oxide, and the rest due to industrial restructuring. Carbon emissions have continued to rise from our homes, cars and services as we continue to consume fossil fuels (EuroStats, 2010).

So achieving a low-carbon society – that is to reduce our CO_2 emissions by at least 90 per cent – in the next 40 years requires drastic changes in either our energy supply or our lifestyles, or preferably a combination of both. Lifestyle changes, centred on improvements in home efficiency and cutting down on carbon-intensive products, like meat, can help. Research presented in Chapter 4 shows that it is possible to reduce CO_2 emissions by nearly 30 per cent by adopting such a modest lifestyle, but the drawback is travel (or mobility) restrictions: no cars,

1

no planes. Just public transport and modest seaside holidays: a return to the 1950s perhaps? So should we be downcast at such a prospect?

As the energy analyst Vaclav Smil (2003: 338) so eloquently wrote some years ago on the prospects of lifestyle changes:

> Such reductions would call for nothing more than a return to levels that prevailed just a decade or no more than a generation ago. How could one even use the term *sacrifice* in this connection? Did we live so unbearably 10 or 30 years ago that the return to those consumption levels cannot be even publicly contemplated by serious policy makers because they feel, I fear correctly, that the public would find such a suggestion unthinkable and utterly unacceptable?

So leaving aside, for the moment, the question of whether a reduced, or different, lifestyle would be acceptable to most people, that still means the bulk of CO_2 reductions must come from changes to our energy system. That is moving away from fossil fuels to low-carbon sources, such as renewables and nuclear power. This book takes a disinterested view in nuclear power; it will still be around in 2050 and important in some countries such as France. But for Britain, our vision is that it will only play a minor role, most probably less important than it is today. This is mainly due to its economic cost – so expensive and risky to build – and the shift to a more decentralised energy system brought about by the extensive use of renewables, particular solar PV and wind. There will be much more local generation and less reliance on the grid for most households and small businesses. Large industrial centres and big cities will, of course, still depend on large power plants, perhaps district CHP or fed by the high-voltage grid from a remote power station (a role for nuclear or coal with carbon capture).

However, this book is not about the technical or economic feasibility of renewable energy systems. It is about trying to imagine what sort of political and social changes would be needed to live in a low-carbon society. We take it for granted that there will be some combination of low-carbon energy systems. The key questions are about the level of energy consumption, and its reliability. The more energy we wish to consume, the higher the level of capital investment. The more reliable the supply, the more we have to invest in storage systems and grid connections. To replicate our current electricity standards of unlimited power on demand, at a constant price with over 99 per cent reliability, could be prohibitively expensive with renewable sources. Our current high-energy lifestyle may not be sustainable in a low-carbon world.

We may choose to adapt to a world of energy restrictions and seasonal availability. Just as once strawberries were only available in the summer, so electric car travel may only be affordable during the summer months. Just as some wish to have a local foods policy, so perhaps we will have a local energy policy consuming only from local suppliers according to the seasons.

Why a low-carbon society?

This book is driven by the belief that we need to move from an energy system based on fossil fuels to one based on low-carbon sources. This is for three reasons: climate change, peak oil and energy security. The first of these, fear of climate change, has been widely accepted by governments and is the rationale for policy measures to support low-carbon sources, such as subsidies for wind and solar energy. The second, peak oil or the belief that oil and gas supplies will soon start to decline, is less accepted. However, current high oil prices (of over $100 a barrel) are an incentive to develop alternative transport fuels, such as biofuels, and the electric car. Finally energy security issues have always bedevilled oil and gas supply from unstable (or undesirable) regimes to Western markets, and have hastened the search for local alternatives.

It is by no means axiomatic that the future of energy supply will be low-carbon; it could be based on abundant and cheap coal (with or without carbon capture), or plentiful (shale) gas. Undoubtedly the average carbon intensity of fuels will continue to fall, as it has done for a century but this does not mean that total carbon emissions will decline. The rate of consumption has always exceeded that of carbon-intensity decline. So it would be unrealistic to expect a reliance on 'business as usual' free-market policies to produce the required cut in emissions. This is a task for governments, driven by strong public pressure.

It would also be unrealistic to expect the world to remain the largely peaceful place it has been for the last 60 years, without any regional wars to disrupt the vast growth in global trade in energy. There are worrying developments for energy security, in piracy, in terrorist attacks and revolts against autocratic regimes. Perhaps voters may come to think that the 'blood and treasure' spent on defending energy supplies might be better spent on subsidising local generation.

Finally, we do not want to give the impression that a low-carbon society is the route to a utopian society – often visualised (as in a short story in Chapter 12) as small rural communities simply and happily existing on solar power, organic farming and crafts. It is not the technology of energy supply that makes people happy but the society that they construct with the technology available. A low-carbon society is no guarantor of personal happiness or a free society. It may just be the best option available.

Telling stories

There are many possible visions for Britain or the world in 2050. To attempt to give some intellectual rigour and credibility to what is basically imaginative speculation on the future, a variety of techniques are used, especially modelling and scenarios. Modelling uses quantitative and numerical analysis while scenarios uses qualitative and literary methods. The former asks 'how much', the latter 'how come'. Both are essential for investigating a low-carbon society. Models can answer the (technical) question of how much wind power (and how many turbines) would be needed, while scenarios could answer the (social and political) question of how are we going to afford them (see, for instance, Skea et al., 2010 for modelling results).

Both techniques are only as good (or as logical) as the assumptions made; they are a stimulus to further thought rather than an accurate description of what will occur. Any scenario might occur, but the probability of it occurring is low. Thus a wide range of scenarios is useful, so that common features or problems may emerge. In this book we have pushed the scenario technique to its (academic) limit by having four short stories about living in a low-carbon society. They range from mildly utopian to dystopian, but what makes happy lives is the structure of society and the opportunities it offers, not the energy choices. As Roger Levett remarks in his story (in Chapter 13):

> [H]ow people behave, and what they think and believe – about the reality of environmental limits and how we should respond to them, about what constitutes the 'good life' ... will be just as significant for whether we make an effective transition to lower carbon living as technical matters.

Thus the starting point for all our scenarios is that (due to some shock event) the public have become sufficiently convinced of the reality and gravity of climate change to make it politically possible for a government to take effective action. What will be different will not be people's aspirations but both the practicalities and the image of low-carbon living. Our current high-consumption lifestyles will then have fallen out of fashion, and the suburban house with three cars will come to be regarded as unfashionable as well as becoming inconvenient to live in. The inner city low-carbon terrace house may then be the ideal home. Levett's story is concerned not with the utopian question of whether a young couple can buy their perfect dream house, but the more realistic – and interesting – one of whether they can get somewhere to live that meets all their basic needs, and that they feel generally comfortable and content with.

As always we will have to live within limits, set by our society and our abilities. Our aspirations will not have changed, neither will our search for happiness and contentment. Thus a low-carbon society is not a utopian society, but just another stage in our long historical journey of using available energy sources to achieve a better life.

Problems of scale

Energy sources are valuable commodities and have to be protected against theft or disruption. To rely upon distant sources means having to protect both the source and the supply line – the electric grid, the gas pipeline, the oil tanker or the coal train. Disruption by some natural event (like flooding or earthquake) or by human intervention (sabotage, piracy or war) causes problems. The solution is to always have multiple sources widely dispersed, as is our global web of suppliers and consumers.

The electricity supply system is particularly vulnerable, grids are easy to disrupt and electricity is hard to store. And a modern life without electricity is very inconvenient, if not impossible. So which is better, a centralised supply with grid or a decentralised one with local storage? Grids are very successful in well-run, ordered societies, not so good in societies suffering from internal conflict. While fossil fuels are often concentrated in certain locations, the sun shines and the wind blows nearly everywhere. Power availability and levels may not be as good using local generation, from wind and solar, as with a centralised grid system, but is it safer or more resilient? Which is it better to pay for, protection of the grid or to invest in local generation?

There are plans for a super grid connecting North African solar PV stations with Europe (see DESERTEC). All very feasible in a peaceful world, but how much are we willing to pay to defend it? The same question, of course, has long existed with oil and gas supplies. So geopolitical issues of what and who to defend will continue to dominate fossil fuel supply. The answers may be very different when oil is $100+ a barrel, rather than $10. But with renewables, there is far less need to defend 'national interests' and consumption will be closely matched to generation, which will require lifestyle changes. In a country like Britain with very variable weather conditions local power levels will fluctuate widely. The key will be electric storage. And the dilemma will be: is it better to invest in (long distance) grid connections or in local storage, or to put up with rationing and occasional blackouts? The answer will depend on the vulnerability of grid systems and the structure of society. These sort of questions are explored in the story by Horace Herring in Chapter 15.

Problems of society

The problems caused by any rapid climate change will have a bigger impact on poorer, more agricultural countries than modern industrial societies. Changing weather patterns – bringing drought or floods – will impoverish poor peasants and cause mass migration. These 'climate refugees' will add to the mass of economic migrants seeking a better life in Western countries. Thus climate change will intensify immigration issues, particularly in southern Europe, and the effects will ripple into northern Europe including Britain.

How we deal with this humanitarian and economic problem will be a test of Europe's liberal immigration and current multicultural policies, which are already under strain. Is a Fortress Europe (or USA) policy of building (physical and legal) barriers against immigration economically feasible and politically acceptable? If violence is used against immigrants, should we not expect retaliation?

These are the themes of the story by Tanya Hawkes (in Chapter 14): Britain's willingness to accept climate refugees and the political divisions it causes in building a low-carbon society. Her lesson is that if countries turn their back on immigrants, they may also be tempted to turn their back on international agreements to control carbon emissions. An insular policy is no solution to global problems, nor is it in the long term any solution to national problems.

So there is always a conflict between the large and the small scale, the local and the national, the decentralised and the centralised. What is best depends on the structure of society, and on the willingness of local communities to integrate and share. Often conflict makes this impossible and communities retreat behind walls into self-sufficiency. Peace makes the dismantling of walls, and the grid possible. But whatever the future, a 'one world' society powered by a global grid or a broken one fractured into many pieces, the sun will shine and the wind will blow on both equally, thus ensuring that they could both be low-carbon societies.

Overview of the book

The book is divided into three parts concerned with policy, case studies and stories. The contents of each are briefly summarised below.

Part I: Policy

This part has five chapters that examine the need, methods and policy required to achieve a low-carbon society. Chapter 2 by Stephen Peake expands the themes of this introduction by asking why we want a low-carbon society, and what it might consist of. Beyond the fact that it will not emit large quantities of carbon dioxide and other anthropogenic greenhouse gases into the atmosphere, there is widespread disagreement on its various technical, behavioural and economic characteristics. As he points out, a low-carbon society is not necessarily a low-energy society or even a low-fossil fuel society. This chapter outlines the political, economic and scientific forces moving us towards a low-carbon society and the policies enacted by the UK government. These are extensive and ambitious, but like the idea of its policy of 'zero-carbon' housing, still the subject of intense debate and scrutiny. There are plenty of visions but, as yet, no consensus as to where we are heading.

Chapter 3 by Nicola Hole gives a detailed account of UK government policies in the residential sector. This sector accounts for about one-third of national carbon emissions, and could cut its emissions by 60 per cent by 2050; two-thirds coming from demand reduction and one-third from low-carbon technologies. Of crucial importance will be behaviour change by consumers and this chapter explores how energy practices or 'habits' are embedded within everyday lifestyles. The key question for government policy is how far current behaviour can be changed towards a low-carbon lifestyle.

Chapter 4 by Angela Druckman and Tim Jackson asks 'what does it mean to live well in a low-carbon society?' They argue that a low-carbon lifestyle must not only deliver the material provisions necessary for an adequate level of nutrition and physiological health but also meet social and psychological needs, for respect and participation in society. Drawing on the work of Townsend and the concept of a 'minimum income standard', they calculate the carbon emissions of the basket of expenditures deemed necessary to enjoy a decent life. Such a lifestyle would reduce household emissions by around a third, but would require giving up much of our current 'status-driven consumerism', such as cars and foreign holidays.

Chapter 5 by Stephen Potter looks at transport, which is the fastest-growing source of CO_2 emissions in the UK, and is possibly the most problematic area with regard to achieving a low-carbon society. This chapter outlines the technical options and policy proposals required for a 60 per cent (or more) cut in carbon emissions for all transport by 2050. Currently, there is strong emphasis on, and interest in, 'low carbon' vehicles (including battery-electric, hybrid electric and fuel cells), and a range of related fuels (bioethanol, biodiesel and hydrogen). However, these technical measures alone are likely to be ineffective, even combined with a substantial shift to public transport what is required is tackling people's need (or desire) to travel, particularly by air.

Chapter 6 by Neil Strachan and Timothy Foxon examines the strengths and limitations of three approaches that policymakers use to explore a transition to a low-carbon society – energy-economic modelling, scenario development and transition pathways. While energy-economic modelling, such as MARKAL, can provide analytical rigour and quantitative results, it is generally not able to deal so well with non-marginal changes. In contrast scenario analysis brings a greater freedom to investigate alternate worlds. Transition pathways are a more recent approach, which focuses on how pathways are influenced by choices by actors. All these techniques show that a wide range of low-carbon futures are feasible, depending on choices made by governments, businesses, community groups and individuals in the next few years.

Part II: Case studies

This part has five chapters that give case studies and present scenarios of low-carbon living. Chapter 7 by Robin Roy presents six case studies of low-carbon lifestyles by pioneering households, small communities and housing providers in Britain. Two are of households that have retrofitted their old homes, two are of new build, and the final two are of low-carbon community housing projects. All achieved substantial savings in energy bills through the use of extensive insulation, passive solar heating plus low-energy lights and appliances, as well as savings in water use. However, they all found it difficult to reduce transport use, and to forgo air travel. Hence living in a low-carbon home is no guarantee of living a low-carbon lifestyle, especially if the householders are relatively wealthy.

Chapter 8 by Catherine Mitchell is her account of converting an old run-down cottage into a low-carbon home. During this process she experiences firsthand the fraught and time-consuming process that many more homeowners could go through if there is to be an ambitious programme of low-carbon retrofitting in the UK, such as through the government's Green Deal programme. Despite the many hassles of being a pioneer, she is very pleased with the end result – a practical, light and comfortable home – with energy bills being reduced by over 80 per cent.

Chapter 9 by Grace Crabb and Adam Thorogood explores the possible effect of climate change on agriculture in Britain. Agriculture now accounts for nine per cent of the UK's greenhouse gas emissions, especially of methane from livestock and nitrous oxide through crop production. However, farming could be carbon neutral or even negative, if there was a return to more traditional methods and an emphasis on soil management to absorb carbon. They present two scenarios in 2050, one for a lowland arable farm and the other for an upland farm in mid-Wales. The major change will be a large reduction of livestock, particularly of sheep and cows, which will mean a major shift in national diets away from meat and dairy consumption to a more vegetarian one.

Chapter 10 by Tanya Hawkes asks what kind of society would willingly embrace rapid decarbonisation. Drawing on the report *zerocarbonbritain2030* by the Centre for Alternative Technology, she argues the case for democracy, for a renewal of the welfare state, and for environmental justice at the heart of international climate change agreements. Her emphasis is on the political steps required to achieve international agreement, through the UNFCCC process, and she supports the creation of an International Court of the Environment to enforce agreements. She argues that a low-carbon society can only be achieved in Britain through strong state intervention, which can set carbon prices and carbon budgets, and ensure sufficient investment in renewables.

Chapter 11 is by Bastien Girod, a green Swiss MP, who outlines the measures needed to achieve a low-carbon society in Switzerland, and the political compromises necessary to achieve them. The three major measures required are zero-emission buildings, low-emission transportation and renewable electricity supply. These are already available today, but the major problem is ensuring long-term public support for a low-carbon society, especially if it involves higher taxes. Thus he argues that what may be deemed most cost-effective by policymakers, such as a carbon tax, may be the least popular. Voters, it seems, prefer subsidies where the benefits are visible but the costs are hidden (in the form of higher electricity prices). Hence the popularity of the domestic feed-in tariffs for electricity produced from wind and solar, and grants for insulation.

Part III: Stories

This part presents four stories about living in a low-carbon society. The use of stories, with characters and dialogue, is unusual in the energy studies but is a long-established technique in some academic fields (Nash, 1990) and is commonly used in organisational change (Gabriel, 2000; Brown et al., 2005) and leadership studies (Owen, 2004). Scenarios are stories about the future, albeit without characters. However, the groundbreaking report, 'Intelligent Infrastructures Futures', had small stories with characters describing their lives and amplifying the dilemmas of the scenarios (OST, 2006). As previous chapters have made clear, people's feelings and attitudes about a low-carbon society are crucial to its realisation, and the exploration of this is the territory of fiction. The first two stories are optimistic, the required changes have not been too painful, while the final two take place in societies racked by social conflict, one caused by climate change, the other by social breakdown.

The first story by James Goodman is an extract from a Forum for the Future Report (FfF, 2006). It describes a fictional village, Little Greenham, in 2016, which has successfully made the transition to a low-carbon society. The second story, 'The Housing Ladder' by Roger Levett, describes the deliberations of an aspiring urban couple to find a new home, and illustrates how fashions and tastes in housing can completely change over time. The third story, 'The Refugee' by Tanya Hawkes, is the memoirs of a climate change refugee, who has been forced to abandon his Pacific island home and comes as a young child to London. It takes place in a Britain struggling socially with mass immigration, and economically with the international pressure to reduce carbon emissions. Its theme is the importance of taking an ethical position on climate change, rooted in equity and fairness. The final story, 'The Gun and the Sun' by Horace Herring, takes place in a Britain fractured by social conflict and energy shortages (a world similar to the scenario 'Tribal Trading' in 'Intelligent Infrastructures Futures'). The story is set in a small self-sufficient village and its theme is, in the absence of any grid, the necessity of adapting to seasonal energy use.

Conclusion

In politics 40 years is a long time. The National Union of Mineworkers and the UK Atomic Energy Authority were the most powerful energy institutions 40 years ago. Both were laid low by a strong-willed prime minister, Margaret Thatcher, who wanted to reshape the energy landscape of Britain. In the next 40 years, such a transformation could once again be seen, if there was another radical prime minister determined to see a low-carbon Britain.

As oil and gas run out, we can expect increasing regional conflicts over access to them. Rather than being drawn into 'oil wars', perhaps it would be better to turn our attention to our renewable energy riches as they are freely available everywhere, and unlike fossil fuels not concentrated in certain locations. But to harness them we need strong government leadership backed by people's willingness to pay for, and adapt to, a low-carbon society. The (short-term) cost is moderate but the (long-term) rewards are immense.

References

DESERTEC Foundation, www.desertec.org.

EuroStats (2010) 'Using Official Statistics to Calculate Greenhouse Gas Emissions', 2010 edition, http://epp.eurostat.ec.europa.eu/cache/ITY_OFFPUB/KS-31-09-272/EN/KS-31-09-272-EN.PDF.

Brown, J., S. Denning, K. Groh and L. Prusak (2005) *Storytelling in Organizations: Why Storytelling is Transforming 21st Century Organizations and Management*, Oxford: Elsevier.

FfF (2006) *Calor Village of the Year Awards: Towards 2016*, London, Forum for the Future, www.forumforthefuture.org/library/calor-village-of-the-year-awards.

Gabriel, Y. (2000) *Storytelling in Organizations: Facts, Fictions, and Fantasies*, Oxford: Oxford University Press.

Nash, C. (1990) *Narrative in Culture: The Uses of Storytelling in the Sciences, Philosophy*, London: Routledge.

OST (2006) 'Intelligent Infrastructures Futures: The scenarios – towards 2055', London, Office of Science and Technology, www.bis.gov.uk/assets/bispartners/foresight/docs/intelligent-infrastructure-systems/the-scenarios-2055.pdf.

Owen, N. (2004) *More Magic of Metaphor: Stories for Leaders, Influencers and Motivators*, Carmarthen; Norwalk, CT: Crown House Pub.

Skea, J., P. Ekins and M. Winskel (eds) (2010) *Energy 2050 – Making the Transition to a Secure Low Carbon Energy System*, London: Earthscan.

Smil, V. (2003) *Energy at the Crossroads*, Cambridge, MA: MIT Press.

Part I
Policy

2
What is a Low-Carbon Society?

Stephen Peake

All visions of a low-carbon society are based on the fact that one way or another they do not emit large quantities of carbon dioxide and other anthropogenic greenhouse gases into the atmosphere, as do so many developed and developing societies today. However, beyond that, such visions differ markedly in their various technical, behavioural and economic characteristics. A low-carbon society is not necessarily a low-energy society, nor even a low-fossil fuel society. For example, many low-carbon scenarios see widespread deployments of coal-fired power stations using carbon capture and storage technologies as vital planks in the transition to a low-carbon future, as well as the use of large-scale renewables and nuclear power.

What is a low-carbon society?

A low-carbon society is the sum of the characteristics of the communities that it comprises. Communities within a low-carbon society may differ in their level of self-sufficiency. Some see communities that are highly integrated and centrally coordinated by a hi-tech super-grid covering a large area (e.g. from Northern to Southern Europe or beyond to North Africa) where the low-carbon energy supplies (renewable, fossil with CCS, nuclear) are balanced, exploiting regional weather patterns and other regional circumstances. Others see a low-carbon society as the sum of a set of essentially-low energy communities, each more or less self-sufficient. Visions of a low-carbon society blend different characteristics such as in Figure 2.1.

- Degree of self-sufficiency versus integrated networks, storage, centralisation.

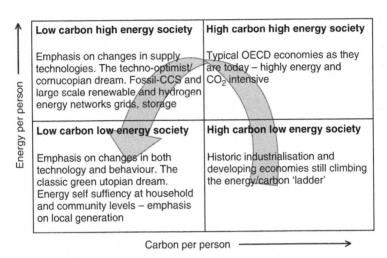

Figure 2.1   The direction of evolution of carbon and energy in society
Source: Author.

- Degree of reliance upon fossil fuel resources.
- Degree of large-scale deployment of renewable, nuclear or hydrogen-based energy systems.

Could there be such a thing as a low-carbon/high income society?

To some extent, the use of energy in the figure above could be interpreted as a proxy for visions of income/consumption (high energy = high consumption).

> The future is already here – it's just not evenly distributed.
> William Gibson, quoted in *The Economist*, 4 December 2003

With our economists hats on we might take inspiration from Gibson's observation, and cast our eyes over international comparisons of carbon productivity looking for the outliers – those countries with unusually high incomes and relatively low-carbon emissions. And we would ask questions about why they were there – what was their secret? We would hope it was not just down to some natural renewable resource endowment or other unique national circumstances that could not be replicated. We might look deeper, beyond those factors and attempt to discern the signature of policy or technology on these low-carbon outliers. Beyond that, we might conjecture something about national cultural/social/lifestyle traits. In fact we've been doing this for years in the literature on international patterns of energy use. However, the mainstreaming of risks associated with climate change and peak oil are forcing us to go beyond mere observation of factors associated with carbon productivity and to devise scientific policies and plans to achieve prescribed decarbonisation pathways.

There are different views on what the historic or cross-sectional international evidence suggests about the relationship between carbon and economic activity – about the determinants of carbon productivity. The lens of the resource productivity (e.g. GDP per unit carbon emitted or its converse resource intensity – carbon per unit GDP) is now the mainstream voice of ecological modernisation. Influential reports call for a 10-fold increase in carbon productivity (economic output per unit carbon emissions):

> Meeting commonly discussed abatement target would require a per-person carbon budget of six kilograms of CO2e per day. If one had to live on such a carbon budget with today's low levels of carbon productivity, one would be forced to choose between a forty kilometer car ride, a day of air conditioning, buying two new T-shirts (without driving to the shop), or eating two meals. So without a major boost in carbon productivity, stabilizing greenhouse-gas emissions would require a major drop in lifestyle for developed countries and would hinder economic development in low-income countries.
>
> (McKinsey, 2008)

Another lens takes a more critical view of the relationship between income and productivity. The controversy over the Environmental Kusnetz Curve as applied to energy or carbon emissions is an example. A classical interpretation of resource productivity suggests that, for example, Bangladesh has half the energy productivity of the UK. However, when this is corrected for income they are very similar (Steinberger and Krausmann, 2011). The critique of carbon productivity is that efficiency improvements occur at a slower rate than absolute increases in consumption brought about by income effects. In other words, getting rich to get green is not necessarily a recipe to live within environmental limits – especially when the clock is ticking and the limits are potentially very close.

Why do we want/need a low-carbon society?

There are several inter-related arguments why we want/need a low-carbon society including:

- To reduce the risks of high energy prices and supply disruption to economic growth and political instability.
- To reduce the risks associated with dangerous climate change.
- To capture the potential synergies between policies which promote low-carbon society and other sustainable development objectives. Pursuing these policies can deliver significant economic, social and environmental co-benefits, especially in developing countries (e.g. Skea and Nishioka, 2008).

Energy security and economic risks

A powerful argument for a low-carbon society is an economic one. For much of the twentieth century fossil fuel prices (and oil prices in particular up to 1973) were relatively cheap. Higher oil prices are having a considerable negative effect on prospects for economic growth. Up to a point, oil-exporting countries benefit greatly from higher energy prices in the short-term (in the longer-term high fossil fuel prices have the effect of encouraging investments in alternatives). The macroeconomic impacts of high energy prices on energy importing economies are measurable. Higher energy prices affect the balance of spending between consumers and energy producers (domestic and foreign) and increase cost of inputs to a wide range of good and services. According to Doug Duncan, Fannie Mae Chief Economist (2011):

> The US consumes roughly seven billion barrels of oil each year, so every dollar increase in the cost of energy translates into a $7 billion annual tax on consumers, reducing their disposable income available for spending on other goods and services.

As recently as last year, the $100 dollar per barrel mark was seen as a pivotal figure – at this point the price transforms the economics of alternatives. A recent DECC study calculated that a doubling of average oil prices from $80 to $160 per barrel would lead to a 1.5 per cent reduction in GDP per year (around £20 billion) (*Guardian*, 2011). Energy imports have major macroeconomic effects. This is because there are myriad connections between fuel prices and other commodities – noticeably food – as seen in the outbreak of food riots in the 2008, the year that oil prices touched over $140 per barrel.

Climate change

Ice and mud cores show that for much of our existence we, *homo sapiens*, had little measurable impact on the carbon, methane and nitrogen cycles of life on earth. The long slow burn of the co-evolution of civilisation and technology was the platform for a steadily increasing (and migrating) population. And then came the industrial revolution and boom! The rest is (recent) history. Humans are dumping carbon dioxide, methane, nitrous oxide and other greenhouse gases at a rate that far exceeds any of the earth's natural capacities to adapt and accommodate. What is more, the rate is increasing as human population increases and its wealth and consumption increases. We have perturbed the carbon cycle at a rate that is a hundred times faster than even the most rapid 'natural example' of rebound from an ice stage of glaciation.

The international community responded in 1992 with the Earth Summit and the launch of Rio Agreements – with a convention on climate change to stabilise greenhouse gas concentrations. At the Copenhagen Climate Summit in 2009 we put a number on that, 2°C, and the science literature responded with an estimation of what this means in terms of low or high carbon pathways. Our best guess is that to have a 50 per cent chance of staying below 2°C increases (relative to the pre-industrial) we need to limit the sum total of historic and future cumulative carbon emissions to one trillion tonnes (Rogelj et al. 2009). Our cumulative historic emissions are half-way there at about 500 GtC. And at our current rate of emissions, we have just around 40 years left before our chances of staying below the 2°C threshold would dip below 50 per cent.

Failure to reduce carbon emissions significantly and dramatically – to improve carbon productivity – will result in a much higher increase in average global mean temperature. Our knowledge of what increases in regional and global temperature means is highly uncertain. Warming that goes much beyond 2°C average increase in global mean surface temperature is thought to lead to substantial economic losses, human health impacts, biodiversity losses but also increase the chances of irreversible 'tipping elements' (such as the slowing down or collapse of the North Atlantic circulation, the collapse of the Amazon basin ecosystem; or the melting of the Greenland ice sheet).

The climate change argument for a low-carbon society is rapidly transforming itself from a 'there and then' argument to a 'here and now' argument. It is not just about 'our grandchildren', there is some evidence that many people alive today are already adversely affected by global warming, and if they are not right now, they soon will be.

Synergies between low-carbon society and sustainability

An ultra low-carbon society (massive increases in carbon productivity) has major implications for materialisation. There is frequently an assumption among some activists and commentators that a low-carbon society is a dematerialised society, where there are other ancillary benefits – less waste, toxic emissions and perhaps more co-benefits (human health improvements through walking and cycling, less cancers from lower carbon diets). The wish list goes on. Deforestation rates are slowed down or halted, and ecological impacts from intensive farming are ameliorated. The key question is whether or not the emerging middle classes in rapidly developing countries such as China and India move towards carbon footprints of the old world economies (Europe) or new world economies (US, Australia).

UK government action

The UK policy landscape around energy conservation, energy efficiency, alternative and renewable energy has developed sporadically since the 1970s. Electricity privatisation and the subsequent introduction of Non Fossil Fuel Obligation in 1990 are the foundations of policies towards low-carbon electricity generation. In the 1990s, domestic efforts to reduce carbon emissions were initially driven by international action under its commitments to the UNFCCC. Government action to reduce carbon emissions was boosted in 2008 when the UK became the first country to adopt legally binding long-term national climate change targets with the passing of the 2008 Climate Change Act. The act made it legally binding for successive ministers to ensure that the 'net UK carbon account for the year 2050 is at least eighty per cent lower than the 1990 baseline'. In 2009, the government published its *National Strategy for Climate and Energy*. The strategy plots how the UK will meet the 34 per cent cut in emissions on 1990 levels by 2020 (this is the 2050 trajectory). The present low-carbon policy landscape is now relatively mature – even crowded) and includes the following policy initiatives designed to assist the transition to a low-carbon economy:

- Carbon Reduction Commitment
- Climate Change Agreements
- Building regulations

- Climate Change Levy
- Renewable Energy Strategy
- EU Emissions Trading Scheme

Although the strategy involves a great many small sectoral changes, in practice the UK's strategy to 2020 is based on five main system changes (savings relative to baseline):

- 250 $MtCO_2$ reductions from power sector and heavy industry
- 50 $MtCO_2$ from policies to reduce CO_2 from new vehicles
- Emphasis in this plan is the following: [categorise where the main savings are coming from]
- 50 $MtCO_2$ from energy efficiency, smart metering, Community Energy Saving Programme, and zero-carbon homes
- 30 $MtCO_2$ from the renewable heat incentive in business and industry
- 20 $MtCO_2$ from changes in agricultural practices.

This is on top of 23-33 $MtCO_2$ savings from measures announced in the 2007 Energy White Paper (many of them listed above).

Global action

While the Copenhagen Climate Summit at the end of 2009 was widely reported and portrayed as a political failure, at the technical/operational level the UNFCCC regime continues to grow in its technical and legal maturity. A 'triple whammy' of Copenhagen, 'Climate Gate' and the 'IPCC glacier debacle' has sufficiently undermined public, business and government confidence in next steps for the international process. Criticism of the failure of the regime (e.g. Prins et al., 2010) to make any real impact on the ultimate goal of the UNFCCC (stabilising greenhouse gas concentrations) needs to be balanced with the institutional capacity that has been built and the learning curve (for example, with carbon projects, carbon markets and domestic programmes of climate mitigation and adaptation).

At the heart of the international response to climate action is an implicit change model that is top down. The premise is that legally binding regimes are necessary to inspire change at national, regional and local levels. The US for example remains outside the Kyoto regime – and we have witnessed local and regional climate action plans (e.g. the C40 cities initiative) over time, as well as attempts at national legislation (e.g. the death of the Waxman-Markey Bill in the US Senate). The European Emissions Trading regime is able and capable of functioning, at least for a period, without immediate international agreement on a successor to the Kyoto Protocol. It is therefore not necessarily the case that slower progress at the international level means less progress in creating and implementing strategies for a transition to a low-carbon society – in some cases international action helps, in others paralysis is itself a source of inspiration for community level action.

Research

The emphasis in current research on low-carbon society seems to be focused on economics, scenario exercises and especially low-carbon *technologies*. Research on low-carbon communities is more limited but emerging. Over the years, the policy agenda has been directed towards an evolving balance of government, business, communities and individuals (Fudge and Peters, 2011). In general there was an emphasis in the 1990s and early 2000s on the individual. A totemic example of this is perhaps the emerging debate around Tradeable Energy Quotas (TEQs) – formerly called domestic tradeable quotas (Fleming and Chamberlin, 2011).

The post-Kyoto emphasis on individualism/consumerism is now being supplemented with new thinking around community. There is an emerging 'community and climate sector' focussing on the role of communities in 'delivering' reduction targets and 'tapping the energy within communities'. One of the early findings seems to be that delivering outcomes within communities by communities can be a slow process. In the UK the Department of Energy and Climate Change has selected 22 'test-bed' low-carbon communities to learn more about the role of communities and community support packages in delivering carbon budgets and renewable energy targets (DECC, 2010). Key elements of the winning bids for this project are shown in Figure 2.2.

There is a great deal of emphasis on energy efficiency and renewable energy retrofit, but there is also strands involving schools and community buildings. Many of the communities plan to re-invest any profits from the government's cash-back scheme into community funds to allow the wider roll out of more community initiatives. Low-carbon living requires transformational changes to systems, technologies and behaviour. The emphasis in these test-beds is clearly on incremental changes through technology, though there are elements of stages of transformation beyond this.

How does the quest for low-carbon society fit in with other environmental debates?

In the post-Copenhagen era we have reached a moment where a pure emphasis on climate and carbon in particular is considered 'so

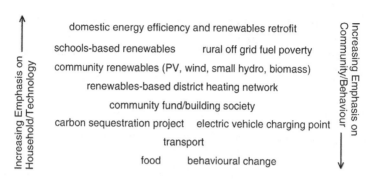

Figure 2.2 Summary of 22 'test-bed' low-carbon community projects in the UK

noughties'. Instead there is now greater emphasis on the complex interplay between a number of different and connected environmental agendas. If a low-carbon society is the answer, what was the problem? Climate change is the obvious problem – but low-carbon living is associated as the solution to other problems (the synergies argument noted above). Three quite different perspectives are helpful in locating low-carbon society within broader environmental debates (Figure 2.3). So-called Growthists, Peakists and Environmentalists all have different takes on the matter (Peake et al., 2011).

A pure 'growthist' perspective emphasises that there will continue to be a very large unmet demand for commercial energy (and other forms of material consumption) for the rest of this century. In the forefront of their minds are the facts that the world's population will increase by another two billion or so by 2050 and that average incomes by then will have risen fourfold. There is an implicit assumption here of a link between consumption and income – indeed the models that generate such forecasts are hard-wired between income growth and consumption. In this case energy (and sometimes fossil fuels) is proxy for wider notions of the link between growth and material consumption. Growthists who accept climate change tend to be technological cornucopians (or ecological modernists) believing that there are plenty of technological options to address climate change.

A pure 'peakist' view questions how much longer conventional (cheaper) supplies of fossil fuels will continue. They look critically at projections of energy growth (and sometimes carbon emissions under various scenarios). A concern might be the extent to which growth may be limited by rising energy prices.

A pure 'environmentalist' perspective emphasises the ecological limits to growth – the need to limit global temperature increase in particular, but also to other problems such as fresh water, deforestation and food supply. For those environmentalists that accept the possibility of peak fossil fuels, the bright side is that energy prices may rise sufficiently to make alternatives competitive and to prevent much of the remaining fossil carbon from being extracted and released into the atmosphere.

Conclusion: What would life be like living in a low-carbon society

How different would a low-carbon society be compared with today? According to the UK DECC's Low Carbon Communities Challenge (2010), in a low-carbon community 'you might see:

- Households, businesses, clubs and societies, faith and youth groups, public sector organisations like libraries, schools and hospitals – all working together – to reduce energy consumption and produce clean, renewable energy for the community
- Secure energy supplies, produced and managed locally

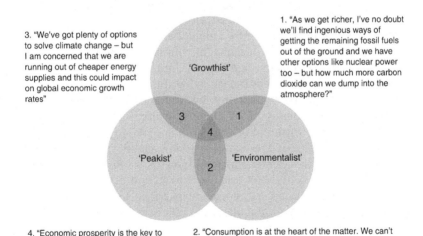

3. "We've got plenty of options to solve climate change – but I am concerned that we are running out of cheaper energy supplies and this could impact on global economic growth rates"

1. "As we get richer, I've no doubt we'll find ingenious ways of getting the remaining fossil fuels out of the ground and we have other options like nuclear power too – but how much more carbon dioxide can we dump into the atmosphere?"

4. "Economic prosperity is the key to solving the twin problems of energy supply and climate change"

2. "Consumption is at the heart of the matter. We can't go on consuming as we do. The bad news is that we are changing the earth's climate. The good news at least is that we may be running out of cheap oil and gas, and possibly coal"

Figure 2.3   Growthist, peakist and environmentalist perspectives on low-carbon societies
Source: Peake et al., 2011.

- Potential to reduce fuel bills through easier ways of keeping homes warm and access to potentially cheaper community owned electricity
- The opportunity to help shape future community approaches to energy issues, testing ideas and providing practical advice and solutions
- Low-carbon jobs created locally
- Food grown locally
- Healthier people who walk and cycle more
- Quieter, safer streets as cars are used less
- People from all parts of the community understanding climate change and the impact of the choices they make and the products they buy on society
- People working together with a sense of pride in what their community is achieving'.

Although how you might 'see' people understanding climate change and the impact of the choices they make and the products they buy on society is not quite clear. There tends to be an emphasis on physical artefacts or processes such as energy technologies, transport or food production in the envisioning of low-carbon communities.

Visions of low-carbon futures are a combination of technical and behavioural. But what kind of lifestyles add up to a low-carbon community? Lifestyle expresses itself at the material, psychological, social and cultural levels (Eyre et al., 2011). So constraining one factor of material lifestyle to 'low carbon' does not necessarily tell us a great deal about the other elements that could go along with that life. Skea and Nishioka (2008) cite the Japan Low-Carbon Society project that envisages a world in which global temperature rise is held below 2°C, global CO_2 emissions are cut by 50 per cent by 2050, and Japanese emissions are cut by 70 per cent. Two visions of a Japanese low-carbon society are contrasted:

> Vision A ('Doraemon') is technology-driven, with citizens placing great emphasis on comfort and convenience. They live urban lifestyles with centralized production systems and GDP per capita growing at about 2% per annum. Vision B ('Satsuki and Mei') is of a slower-paced, nature-oriented society. People tend to live in decentralized communities that are self-sufficient in that both production and consumption are locally based. This society emphasizes social and cultural values rather than individual ambition.
>
> (Skea and Nishioka, 2008: 8)

However, both visions say little about society – they are essentially two different descriptions of technological futures – a high-energy (nuclear) and a low-energy (gas and biomass) future. In a stimulating lecture Walt Patterson (2010) makes a powerful plea for clarity around what we mean by a low-carbon society:

> We get energy services not from fuel alone but from systems – systems of technology, that in turn are created and maintained by social and financial systems – that is, by people working together accordingly to an agreed understanding and a corresponding framework of rules. To manage energy right we urgently have to refine our understanding and refocus the rules. We can start by stating explicitly what we know to be true but have been too mealy-mouthed to say out loud. We talk about a low-carbon UK, a low-carbon future in a low-carbon world. Let's not be coy about this. *Low carbon means low fuel.* We need to use less fuel (emphasis added).

When it comes to envisioning what we mean by low-carbon society, it is by no means clear that we all agree that they are low-energy, or even low-fuel societies. On a very practical level, the UK has a policy that new housing will be 'zero carbon' from 2016. The interpretation of what is meant by 'zero carbon' in this context has been the subject of intense debate and scrutiny. It is anticipated that zero-carbon UK homes will be by no means zero-fuel homes. As we head towards designs for 'zero-carbon' buildings that use fuel – it is equally unclear where our visions of low-carbon communities/society will lead us.

References

DECC (2010) 'The Low Carbon Communities Challenge', Department of Energy and Climate Change, London.

Eyre, N., J. Anable, C. Brand, R. Layberry and N. Strachan (2010) 'The Way we Live from Now On: Lifestyle and Energy Consumption', in J. Skea, P. Ekins and M. Winskel (eds) *Energy 2050 – Making the Transition to a Secure Low Carbon Energy System*, London: Earthscan.

Duncan, D. (2011) 'Economic Outlook', Fannie Mae's Economics & Mortgage Market Analysis Group, Washington, DC, March 2011.

Fleming, D. and S. Chamberlin (2011) 'Tradable Energy Quotas A Policy Framework for Peak Oil and Climate Change', House of Commons, All Party Parliamentary Group on Peak Oil & The Lean Economy Connection.

Fudge, S. and M. Peters (2011) 'The National Dialogue on Behaviour Change in UK Climate Policy: Some Observations on Responsibility, Agency and Political Dimensions', Working Paper 04-11, University of Surrey, RESOLVE.

Guardian (2011) 'UK Facing 1970s-Style Oil Shock which could Cost Economy £45bn – Huhne', *The Guardian*, 3 March 2011, http://www.guardian.co.uk/environment/2011/mar/03/chris-huhne-oil-prices-green-economy.

McKinsey (2008) 'The Carbon Productivity Challenge: Curbing Climate Change and Sustaining Economic Growth', McKinsey Global Institute, www.mckinsey.com/mgi/publications/Carbon_Productivity/index.asp.

Patterson, W. (2010) 'Managing Energy Wrong', Energy Seminar, University of Surrey, 20 January 2010.

Peake, S., B. Everett and G. Boyle (2011) 'Introducing Energy Systems and Sustainability', in B. Everett, G. Boyle, S. Peake and J. Ramage (eds) *Energy Systems and Sustainability*, Oxford: Oxford University Press.

Prins, G., I. Galiana, C. Green, R. Grundmann, M. Hulme, A. Korhola, F. Laird, T. Nordhaus, S. Rayner, D. Sarewitz, M. Shellenberger, N. Stehr and H. Tezuka (2010) 'The Hartwell Paper: A New Direction for Climate Policy after the Crash of 2009', Oxford/LSE, sciencepolicy.colorado.edu/admin/publication_files/resource-2821-2010.15.pdf.

Rogelj, J., B. Hare, J. Nabel, K. Macey, M. Schaeffer, K. Markmann and M. Meinshausen (2009) 'Halfway to Copenhagen, no way to 2°C, Nature Reports Climate Change', Published online: 11 June 2009 | doi:10.1038/climate.2009.57.

Skea, J. and S. Nishioka (2008) 'Policies and Practices for a Low-Carbon Society', *Climate Policy* 8: S5–S16.

Smil, V. (2003), *Energy at the Crossroads*, Cambridge, Mass.: MIT Press.

Steinberger, J. and F. Krausmann (2011) 'Material and Energy Productivity', *Environmental Science & Technology*, 45(4): 1169–76.

3
Low-Carbon Living in 2050

Nicola Hole

The UK is committed to reducing greenhouse gas emissions by 80 per cent by 2050 from a 1990 baseline. In order to reach this target the energy system will need to undergo a 'transition' on all levels, including those relating to human actions and behaviours. Responsible for approximately 30 per cent of total carbon emissions in the UK, the residential sector has a crucial role to play in reaching national CO_2 reduction targets (Palmer et al., 2006). It has been estimated that this sector has the potential to cut emissions by 60 per cent between 1996 and 2050, with two-thirds of the reduction in this scenario coming from demand reduction and one-third from low- and zero-carbon technologies (Boardman et al., 2005). Realising reductions of this scale will undoubtedly involve 'substantial behaviour change from all sectors of society' (HMG, 2009).

Policy approaches in the transition to sustainable lifestyles

While an important component in creating a low-carbon economy will be ensuring that the UK's energy supply is decarbonised, it has more recently been acknowledged that the only way to achieve significant cuts in emissions is through 'ambitious per capita demand reduction' (HMG, 2010a). In combating climate change, increasing the efficiency of energy use represents a cost effective and relatively straightforward step in closing the gap between supply and demand. With at least 80 per cent of the homes standing in 2050 having already been built, there is clearly a strong need to address the efficiency of the existing housing stock, one of the least efficient in Europe (Boardman et al., 2005).

While the technological advancements necessary for meeting both renewable energy and energy efficiency targets appear feasible (CCC,

2008), the government has so far had to rely heavily upon financial incentives and regulation in order to realise emissions savings in these areas. In 2002, the Department of Environment, Food and Rural Affairs (Defra) introduced the Energy Efficiency Commitment (EEC), a three-year energy saving objective for domestic energy suppliers. By the 2004 Housing Act a target was established to improve residential energy efficiency in England by at least 20 per cent by 2010 from a year 2000 baseline. The EEC was renamed the Carbon Emissions Reduction Target (CERT) in 2008 and remains the primary government scheme aimed at increasing the efficiency of the current domestic housing stock. CERT runs as a legally binding obligation placed on gas and electricity suppliers to spend a fixed amount per customer on efficiency measures or renewable technologies each year. Ancillary initiatives in place include Warm Front, designed to deliver energy efficiency measures to those on the lowest incomes, and various activities run through the Energy Saving Trust. Given that space heating makes up the greatest segment of household energy consumption (around 60 per cent), this is the area in which the majority of improvements have been made, via cavity wall and loft insulation and high-efficiency condensing boilers (DECC, 2010a).

Within the Low Carbon Transition Plan (HMG, 2009), released under the Labour government in 2009, proposals were announced for a far more efficient housing stock by 2050, with the majority of heat and electricity to be generated from low-carbon sources. Shortly after the 2010 general election, Prime Minister David Cameron pledged that the coalition government would be the greenest government ever, announcing their flagship Green Deal programme, a key element of their energy strategy (HMG, 2010b). The Green Deal is designed to overcome both the upfront and hidden costs associated with investing in efficiency improvements that are often cited as barriers in preventing consumers from realising energy-saving opportunities (NAO, 2003; Ofgem, 2009). It will offer loans with repayments that are lower than the expected savings and which are tied to properties rather than individuals. Coming into effect in the autumn of 2012, the Green Deal has clear potential to help unlock emissions reductions from homes and commercial properties, but the generation of sufficient demand will be crucial to the success of the programme (CBI, 2011). A significant increase is required in the uptake from older homes which are harder to treat (CCC, 2010), but the extent to which demand for such measures can be stimulated from households is unclear (NAO, 2008). Evidence would suggest that a lack of upfront capital is not the only barrier to action and that less tangible psychological and sociological factors also need to be considered when trying to overcome consumer inertia (NERA, 2007; McNamara and Grubb, 2011). The challenge for the Green Deal will be in creating a mainstream market demand for energy efficiency (UKGBC, 2008).

Savings from efficiency measures have so far been less than optimum, due mainly to a steadily rising number of households, containing an increasing number of energy-demanding appliances (Boardman, 2004). An increase in the number of households has meant that many of the gains made through technological and efficiency improvements have been unable to counterbalance the rising demand for energy. Between 1990 and 2009 there was a nine per cent reduction in energy consumption per UK household and yet, within the same period, overall domestic energy consumption increased by seven per cent (during which time the number of households in the UK increased by 18 per cent) (DECC, 2010b). Despite the array of policy mechanisms in place to encourage the adoption of efficiency measures, uncertainty remains over the extent to which the gains from efficiency improvements can be relied upon to deliver significant reductions in carbon emissions. In countries often thought to set standards on energy efficiency, such as Denmark and the Netherlands, there has not been the fall in emissions that you might have expected considering their aggressive energy-efficiency programmes (Keay, 2005). While the demand for energy services continues to rise, it will remain difficult for efficiency savings to lead to associated reductions in overall energy demand.

The market for domestic energy efficiency within the UK is complex and highly fragmented (UKGBC, 2008), based on various interrelated policies and involving a range of decision-makers and stakeholders. Incremental updates in framework policies instigated by short-term politics will only serve to unhinge a sector reliant on a stable and predictable policy landscape (Brophy Haney et al., 2010). At the household level, more emphasis needs to be placed on tackling homes that are harder to treat, technologies that are newer and more expensive and householders who are less keen to act (NAO, 2008).

Alongside tackling energy efficiency and increasing low-carbon energy generation, reducing energy demand through behaviour change will play a crucial part in the transition to a low-carbon future. The UK Energy Research Centre (UKERC) has estimated that social and lifestyle change has the potential to reduce carbon emissions and national energy use by 30 and 35 per cent below 1990 and 2000 levels respectively (UKERC, 2009). Reducing demand through behavioural change can also help to insure against barriers that may arise in technological deployment, such as failures with, or public resistance against, key supply side technologies.

Targeting behaviour to reduce demand

Policy often seeks to encourage a change away from undesirable conduct, whether related to the environment, human health or anti-social behaviour. There is a long history of enacting policies aimed at altering environmentally sensitive behaviour, ranging from the tax on landfill to the Climate Change Levy. Interventions to try and stimulate a change in behaviour are commonly directed by a basic conceptualisation, based upon some various underlying assumptions about the nature of development and learning, and implicit assumptions regarding human behaviour. Conceptual models have been developed to aid understanding around why people do or do not adopt pro-environmental behaviours, built from a set of conceptual premises, and some form of cause-and-effect relationship (Jackson, 2005). Many influential and commonly used theoretical frameworks have been developed to help ascertain the factors, both positive and negative, that have an influence on pro-environmental behaviour. The effective implementation of policies that alter behaviours relies explicitly or implicitly on these 'models' of what environmental behaviour is, and how it is influenced, shaped and modified by a range of drivers.

The weakness with this strategy is that environmental behaviour is inherently complex, while the models that are guiding behaviour change interventions are, by definition, based upon very simplistic conceptions of why people behave in the way they do. The early linear models of the 1970s epitomise this basic connection, usually implying that an increase in environmental knowledge creates a pro-environmental attitude, which in turn leads to a pro-environmental behaviour. Since then, the research base has progressed to look at novel processes (e.g. subjective norms and perceived behavioural control (Ajzen, 1991), effort (Schultz and Oskamp, 1996) and habits (Verplanken et al., 1994)) to try and understand their role in moderating the attitude–behaviour link. Despite top-level recognition that behaviour is inherently embedded in social situations, institution contexts and cultural norms (DEFRA, 2005), policy continues to frame pro-environmental behaviour change as a straightforward choice that all individuals or businesses are able to make (HMG, 2009: 36, 143, 170). Viewing behaviour change in this way removes responsibility from other actors and sectors that may have an influence on the difficulty or ease with which pro-environmental decisions can be made. This individualistic approach that has been in favour for so long needs to be re-examined in the light of the latest research into low-carbon lifestyles. Channelling policy and programme development towards individualised behaviour change is misguided if it dismisses the broad regulatory, institutional and social setting in which these behaviours form (Moloney et al., 2010), yet there remains a disconnect between government claims regarding the scale and nature of change required on one hand, and what is being invested in through current policy initiatives on the other (Hargreaves and Restorick, 2006).

Alongside personal knowledge, attitudes and values, there are numerous social, cultural and institutional influences that shape the way in which individuals behave, and which can upset the accuracy of individualistic conceptualisations of environmental behaviour. The incremental steps that have resulted from policy campaigns aimed at changing behaviour are likely to be inadequate for producing the manner of social and lifestyle change that is needed to reduce energy use by up to 35 per cent. Borrowing from the transitions literature, a multi-level perspective (MLP) (Rip and Kemp, 1998; Geels, 2002) may be better suited to investigations around the ways in which to engender low-carbon lifestyles. Taking a multi-level perspective presents the

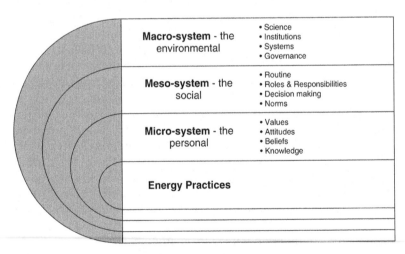

Figure 3.1   A multi-level perspective for addressing the factors that affect energy consumption
Source: Author.

energy system as a series of interlocking and interrelated levels: the micro, macro and meso (Figure 3.1) (Shenk et al., 2007; Rotmans et al., 2003). The micro level of lifestyles relates to the individual; the meso to different social situations, routines, roles and responsibilities; and the macro level of governance, institutions and infrastructure. It becomes apparent from this view, that concentrating all resources on the micro-system without regard to other levels, omits crucial elements in the process of change.

Each level of the energy demand system is positioned to play a unique role in fostering the transition to low-carbon lifestyles; an understanding of each level and the interconnections between them will be an important first step in policy design. Applying this model to social and lifestyle transitions builds a more comprehensive picture of the way in which lifestyles are currently constructed.

From behaviours to practices

In acknowledging the connection between attitudes and behaviour as error-prone and complex (Ajzen and Fishbien, 2005), analysis is increasingly being repositioned to place practices at the centre of enquiry. This 'practice approach' has been adopted across a range of disciplines, including sociology (Warde, 2005; Shove, 2003; Shove and Pantzar, 2005) and anthropology (Evens and Handelman, 2006). In taking the practice as a point of reference, individuals become a part of an interlocking puzzle; no more or less important than the technologies, social processes, cultural frameworks, systems of provision and institutions that surround them. Applying a theory of practice may help to develop a more complete picture of the energy demand system and the way various incumbent actors and systems help to construct current lifestyles and obstruct or unlock new practices. It allows a greater understanding into not only the conditions that surround energy conservation behaviour but also into what is regarded by consumers as normal and over-consumption. This removes the focus both from individuals and from the 'environmental' nature of practices, leading to a more systematic understanding of the embedded nature of current lifestyles and therefore the type and extent of changes that are required to encourage more sustainable lifestyles in the future.

Learning from research into socio-technical systems

The adoption and effective use of different low-carbon technologies is clearly a vital step in developing low-carbon lifestyles (Roy and

Caird, 2006) but inefficient or badly designed products that do not deliver on user requirements will fail to deliver expected emissions reductions and provide little in the way of consumer satisfaction. Lessons learnt from research into low-carbon energy technologies can be utilised within behaviour and lifestyle research. Low-carbon buildings, for example, with innovative technical systems for heating, cooling and lighting, on the surface demonstrate best practice in the challenge to decarbonise the domestic sector. However, a disregard for the context in which a building is to be used and the requirements of those occupying it can render even the most innovative of technologies useless.

Box 3.1   Active House

Active House is a project that is attempting to prevent such gaps between estimated and actual energy reductions. The Active House vision is to produce buildings that create healthier and more comfortable lives for their occupants without negative impact on the environment. To do this, the project attempts to take a balanced and holistic approach to building design and performance, and works on the premise that energy and environmental requirements represent opportunities as opposed to limitations in the creation of buildings fit for a low-carbon future. They propose taking a multi-disciplinary approach; that is taking qualitative and quantitative aspects of low-carbon design into account in the design and valuation of buildings. In trialling new building design, energy consumption is measured alongside technical characteristics (room temperature, CO_2, human presence, window control, lighting, awnings and blinds), as well as so-called liveability factors (the ability to sense the changeability of daylight, the scents and sounds of fresh air). With such valuation, creating an environment that is healthy, comfortable and that satisfies human experience becomes as important as metered energy consumption and carbon emissions. Emphasis is increasingly being placed on the role that designers can play in shaping new paradigms of energy consumption (Pierce et al., 2010), and the responsibility of actors at the design and planning stages to ensure that the human aspects of low-carbon technologies are considered.

Source: Active House (http://www.activehouse.info/).

Applying a new dimension to lifestyle transition research

In the same way that technological advancements and human behaviour cannot be separated (Janda, 2011), nor can the social, cultural and institutional elements be separated from the individual energy-consuming practices that shape lifestyles. Developing a whole systems approach within our understanding of lifestyle transitions and behaviour change will require future research to widen its reach beyond the narrow confines of attitudes and awareness. Progressive policy action that facilitates a transition to truly low-carbon lifestyles calls for an appreciation of the interlocking nature of the micro, meso and macro levels of the demand-side energy system. So for example, a systemic awareness of the energy practices relating to food consumption would involve understanding the micro (tastes, preferences, experience), meso (cooking, storing and shopping for food; household food routines) and macro (establishments, systems, pricing, agricultural policy) drivers of demand.

At the heart of this vertically integrated approach is that emissions reductions need to be juxtaposed with lifestyle value. In the same way that low-carbon buildings need to be designed and considered in a way that satisfy 'user requirements', so too does policy that aims to make current ways of living less energy-intensive. In order to engender the type of lifestyle alterations that will reduce levels of energy consumption by 35 per cent, persistent reliance on environmental and financial benefits of change fails to capture the full range of motives that shape the way practices are undertaken. This point can be illustrated by a look at the food system and the way in which the last 50 years has brought about a major shift in the provision of food. Vaughan et al. (2007) use milk consumption to illustrate the way in which the transportation, storage and refrigeration of milk has shifted from supplier to consumer and in doing so has created new patterns of behaviour (bulk shopping, for example). The evolution of the food system has significantly reduced the time households need to labour over food gathering, preparation and cooking, one factor that has enabled longer working weeks, flexible working hours and more household members in full-time employment. To suggest households return to a more localised and frequent pattern of food shopping disregards the embedded nature of current practices and the technologies, cultures and systems that have built up around them. This is not to say that new systems around food and other practices cannot be developed, rather that it will involve a more comprehensive understanding of the 'vertical system-of-provision' perspective within energy-related consumption behaviour, and the importance of the 'interconnectedness of production and consumption' (Spaargaren and Van Vliet, 2000). Policy initiatives need to be aware of, and if possible address, all the interconnected levels that regulate lifestyles, recognising the embedded nature of present routines and developing alternatives that are plausible, appealing and accessible.

Deconstructing the layers of influence

Deconstructing current energy practices will bring to light the way in which different actions are bound up in the interlocking structure of the current energy system. Although time consuming and labour intensive, developing research that cultivates a more comprehensive and multi-dimensional understanding of contemporary lifestyle situations will lead to more effective and balanced policy action. Analysing the 'whole' energy system from the level of consumption may allow transitional inroads that are far more insightful than viewing, as is so often the case, the functioning of a system from highly aggregated data (Focacci, 2003). Focussing on individual practices and building up a picture of the personal, social and environmental systems that shape them will allow structural mechanisms and the interactions between micro, macro and meso levels to be observed (Haanpaa, 2007). There is a lot still to be unearthed concerning the emergence and subsequent changes in the patterns and performances that structure social life – including how they develop, whether they are borne out of circumstances of systems and technologies and how these change over time (Ensminger and Knight, 1997).

Conclusion

The exact path the UK will take in order to reach the legally binding targets set out in the Climate Change Act of 2008 remains to be seen. A small tipping point for an economy based on low-carbon energy generation was passed in 2008, when global investment in green energy overtook that for fossil-fuel generation (UNEP/NEF, 2009). Although difficult to measure, it is unlikely that this same moment of change in the transition to sustainable low-carbon lifestyles has yet been reached – or is likely to be reached in the very near future. The post-industrial shift from a low to high-carbon economy saw technologies and lifestyles evolve as one and the advantages of this transition sold themselves (improved mobility and communication, for example). Freewheeling in such a way to a low-carbon economy is not an option. Many benefits in taking a low-carbon route once again are long-term and intangible

(POST, 2008) and ensuring the transition will involve significant public policy intervention. This is especially challenging given that carbon emissions are currently coupled closely with economic growth – and so moving to a low-carbon economy implies either decoupling or degrowth (Jackson, 2009).

In October 2010, the government launched a review into the use of behaviour change interventions to achieve policy goals. The outcome of this report will be published during the summer of 2011 and may signal whether or not there will be a genuine move beyond current models of behaviour change intervention and whether a style of policy-making can emerge that acknowledges the way in which the state and other actors configure the fabric and texture of daily life (Shove, 2010). Understanding the landscape and conditions within which individuals consume energy will be an important component in creating a system where production and consumption are more intelligently aligned. Demand reduction and management can play an important role in contributing to the UK's goal for a clean, affordable and secure energy system. Recognising this and ensuring the potential of this role is fulfilled requires a more comprehensive and integrated understanding of how energy practices are embedded within lifestyles.

References

Ajzen, I. (1991) 'The Theory of Planned Behavior', *Organizational Behavior and Human Decision Processes* (50): 179–211.

Ajzen, I. and M. Fishbein (2005) 'The Influence of Attitudes on Behavior', in D. Albaracín, B. T. Johnson, and M. P. Zanna (eds), *The Handbook of Attitudes*, Mahwah, NJ: Lawrence Erlbaum.

Boardman, B. (2004) 'New Directions for Household Energy Efficiency: Evidence from the UK', *Energy Policy*, 32 (17): 1921–33.

Boardman, B., S. Darby, G. Killip, M. Hinnells, C. Jardine, J. Palmer, and G. Sinden (2005) *40% House*, University of Oxford: Environmental Change Institute.

Brophy Haney, A., T. Jamasb, L. M. Platchkov and M. G. Pollitt (2010) 'Demand-side Management Strategies and the Residential Sector: Lessons from International Experience, Cambridge Working Papers in Economics', Faculty of Economics, University of Cambridge.

CCC (2008) 'Building a Low-Carbon Economy – the UK's Contribution to Tackling Climate Change', London: Committee on Climate Change.

CCC (2010) 'Meeting Carbon Budgets – Ensuring a Low Carbon Recovery', Second Progress Report, London: Committee on Climate Change to Parliament.

CBI (2011) 'The Real Deal? Making the Green Deal Work', Confederation of British Industry, http://www.cbi.org.uk/ndbs/press.nsf/0363c1f07c6ca12a8025671 c00381cc7/6e247a796edd217d80257837004f2954/$FILE/CBI%20Green%20D eal%20Feb%2011.pdf.

DECC (2010a) 'Sustainable Energy Report 2010 Progress against the Government's English Household Energy Efficiency Target', London: Department of Energy and Climate Change, http://www.decc.gov.uk/assets/decc/What%20we%20do/Supporting%20consumers/saving_energy/785-sustainable-energy-report-2010.pdf.

DECC (2010b) 'Digest of United Kingdom Energy Statistics 2010', London: Department of Energy and Climate Change.

DEFRA (2005) 'Changing Behaviour through Policy Making', London: Department of Food and Rural Affairs.

Ensminger, J. and J. Knight (1997) Changing Social Norms: Common Property, Bridewealth, and Clan Exogamy, *Current Anthropology*, 38 (1): 1–24.

Evens, T. and D. Handelman (2006) *The Manchester School: Practice and Ethnographic Praxis in Anthropology*, Oxford: Berghahn.

Focacci, A. (2003) 'Empirical Evidence in the Analysis of the Environmental and Energy Policies of a Series of Industrialised Nations, during the Period 1960–1997, using Widely Employed Macroeconomic Indicators', *Energy Policy*, 31: 333–52.

Geels, F. W. (2002) 'Technological Transitions as Evolutionary Reconfiguration Processes: A Multi-Level Perspective and a Case-Study', *Research Policy*, 31 (8/9): 1257–74.

Haanpaa, L. (2007) 'Structures and Mechanisms in Sustainable Consumption Research', *International Journal of Environment and Sustainable Development* 6 (1): 53–66.

Hargreaves, T. and T. Restorick (2006) 'Changing Environmental Behaviour: A Review of Evidence from Global Action Plan', London: Global Action Plan.

HMG (2009) 'The UK Low Carbon Transition Plan: National Strategy for Climate and Energy', London: TSO.

HMG (2010a) '2050 Pathways Analysis', Department of Energy and Climate Change, http://www.decc.gov.uk/assets/decc/What%20we%20do/A%20low%20carbon%20UK/2050/216-2050-pathways-analysis-report.pdf.

HMG (2010b) 'The Coalition: Our Programme for Government', London: HM Government.

Jackson, T. (2005) 'Motivating Sustainable Consumption: A Review of the Evidence on Consumer Behaviour and Behavioural Change', Report to the Sustainable Development Research Network, London: Policy Studies Institute.

Jackson, T. (2009) *Prosperity Without Growth: Economics for a Finite Planet*, London: Earthscan.

Janda, K. (2011) 'Buildings Don't Use Energy: People Do', *Architectural Science Review*, 54: 15–22.

Keay, M. (2005) 'CO2 Emissions Reductions: Time for a Reality Check?' Oxford Institute for Energy Studies Energy Comment February 2005, http://www.physics.harvard.edu/~wilson/energypmp/Reality_Check.pdf.

McNamara, S. and M. Grubb (2011) 'The Psychological Underpinnings of the Consumer Role in Energy Demand and Carbon Abatement', Cambridge Working Papers in Economics No. 1126, Faculty of Economics, University of Cambridge. http://www.econ.cam.ac.uk/dae/repec/cam/pdf/cwpe1126.pdf.

Moloney, S., R. Horne, and J. Fien, J. (2010) 'Transitioning to Low-Carbon Communities – from Behaviour Change to Systemic Change: Lessons from Australia', *Energy Policy*, 38 (12): 7614–7623.

NAO (2003) 'Warm Front: Helping to Combat Fuel Poverty', London: National Audit Office.

NAO (2008) 'Programmes to Reduce Household Energy Consumption', London: National Audit Office.

NERA (2007) 'Evaluation of Supplier Obligation Policy Options: Report for DTI and Defra', NERA Economic Consulting, http://webarchive.nationalarchives. gov.uk/+/http://www.berr.gov.uk/files/file38976.pdf.

Ofgem (2009) 'Can Energy Charges Encourage Energy Efficiency? A Discussion Paper to Prompt Debate', Office of Gas and Electricity Markets, http://www. ofgem.gov.uk/sustainability/Documents1/Final%20discussion%20paper%202 2%20July.pdf.

POST (2008) 'The Transition to a Low Carbon Economy', Parliamentary Office of Science and Technology POSTnote 318, http://www.parliament.uk/documents/ post/postpn318.pdf.

Palmer, J., B. Boardman, C. Bottrill, S. Darby, M. Hinnells, G. Killip, R. Layberry, H. Lovell (2006) 'Reducing the Environmental Impact of Housing', Consultancy study in support of the Royal Commission on Environmental Pollution's 26th Report on the Urban Environment, University of Oxford: Environmental Change Institute.

Rip, A. and R. Kemp (1998) 'Technological change', in S. Rayner and E. Malone (eds), *Human Choice and Climate Change*, Volume 2: Resources and Technology, Washington: Battelle, Press, pp. 32799.

Rotmans, J., J. Grin, J. Schot, and R. Smits (2003) 'Multi-, Inter- and Transdisciplinary Research Program into Transitions and System Innovations', (Maastricht: ICES-KIS-Research Programme) http://www.nido.nu/image/ publicatie/bestand/1048677537.pdf.

Roy, R. and S. Caird (2006) 'Designing Low and Zero Carbon Products and Systems – Adoption, Effective Use and Innovation', Proceeding of *Sustainable Innovation 06: 11th International Conference*, 23–24 Oct 2006, Chicago, USA.

Schenk, N., H. Moll and A. Uiterkamp (2007) 'Meso-Level Analysis, the Missing Link in Energy Strategies', *Energy Policy*, 35: 1505–16.

Schultz, P. and S. Oskamp (1996) 'Effort as a Moderator of the Attitude-Behaviour Relationship: General Environmental Concern and Recycling', *Social Psychology Quarterly*, 59 (4) 375–83.

Shove, E. (2003) *Comfort, Cleanliness and Convenience: The Social Organization of Normality*, Oxford, Berg.

Shove, E. and M. Pantzar (2005) 'Consumers, Producers and Practices: Understanding the Invention and Reinvention of Nordic Walking', *Journal of Consumer Culture*, 5: 43–64.

Shove, E. (2010) 'Beyond the ABC: Climate Change Policy and Theories of Social Change', *Environment and Planning A*, 42:6 (2010): 1273–85.

Spaargaren, G. and B. Van Vliet (2000) 'Lifestyles, Consumption and the Environment: The Ecological Modernisation of Domestic Consumption', *Environmental Politics*, 9, 50–76.

UKGBC (2008) 'Low Carbon Existing Homes', UK Green Building Council, Department of Communities and Local Government, www.ukgbc.org/site/ document/download/?document_id=371.

UKERC (2009) 'Energy 2050 – Making the Transition to a Secure Low-Carbon Energy System', London: UK Energy Research Centre. http://www.ukerc.ac.uk/ Downloads/PDF/U/UKERCEnergy2050/0906UKERC2050.pdf.

UNEP/NEF (2009) 'United Nations Environment Programme, Global Trends in Sustainable Energy Investment 2009: Analysis of Trends and Issues in the Financing of Renewable Energy and Energy Efficiency', Paris: UNEP and New Energy Finance Ltd.

Vaughan, P. and M. Cook (2007) 'A Sociology of Reuse: Deconstructing the Milk Bottle', *Sociologia Ruralis*, 47(2): 120–34.

Verplanken, B., H. Aarts, A. Knippenberg and C. Knippenberg (1994) 'Attitude Versus General Habit: Antecedents of Travel Mode Choice', *Journal of Applied Social Psychology*, 24 (4) 285–300.

Warde, A. (2005) 'Consumption and Theories of Practice', *Journal of Consumer Culture*, 5: 131–53.

4
What is the Carbon Footprint of a Decent Life?

Angela Druckman and Tim Jackson

What does it mean to live well in a low-carbon society? Clearly, the challenges of climate change and resource depletion are not just techno-logical. Behaviours and lifestyles will also need to change. Consumption levels in richer nations look unsustainable when viewed from the perspective of global equity within carbon limits. A world in which a projected nine billion people all aspire to the levels of material comfort expected in the affluent West looks almost impossible to achieve. And yet it is also unrealistic to suppose that people will voluntarily forego the capabilities at least to live a decent life. Sustainable lives must still be worth living.

What it means to live well is always going to be a subject for debate. But several things are clear at the outset. Firstly, it is nonsense to suppose that people can flourish without food, clothing and shelter. Sustainable living must deliver the material provisions necessary for a reasonable level of nutrition and physiological health. But equally it is clear that material needs do not exhaust our appetite for material goods. Much of what we mean by living well can be characterised as social and psychological – rather than material – in nature. Once material needs are met, people's ability to flourish depends on their 'ability to give and receive love, to enjoy the respect of [their] peers, to contribute useful work, and to have a sense of belonging and trust in the community' (Jackson, 2009).

Joseph Rowntree Foundation

In his seminal work on poverty in the UK, Townsend (1979) made a very similar point. Recognising that incomes and expenditures are not a good guide to poverty, he argued that poverty is structural. That is produced by the organisation of society rather than by individual failings, and that inste4ad of using monetary measures, poverty should be measured in terms of relative deprivation. Moreover, he theorised poverty not simply in terms of having enough to satisfy material needs for survival, but in terms of the ability to participate in social activities and to have living standards comparable to current norms.

This multifaceted understanding of poverty – and indeed of the good life – still motivates poverty research today. In a recent study for the Joseph Rowntree Foundation (JRF), Bradshaw et al. (2008) set out to define an acceptable 'minimum income standard' – a level of income needed to achieve a decent life. Specifically, the study attempted to identify an acceptable standard of living that would include 'more than just, food, clothes and shelter. It is about having what you need in order to have the opportunities and choices necessary to participate in society' (Bradshaw et al., 2008: 1).

The JRF study established the basket of expenditures deemed necessary to enjoy a decent life. These were drawn up through consensual discussions among ordinary people, informed at successive stages by feedback from experts. This process established a series of expenditure budgets for a variety of different types of household types (parents with two children, single pensioner households and so on) in the UK. In accordance with the discussion above, the budgets provided for more than adequate warmth, food and shelter. They included significant resources to 'participate in society and maintain dignity' (ibid). On the other hand, these budgets sought to exclude items that might be regarded as 'aspirational', since the minimum income standard is about 'about fulfilling needs and not wants' (Bradshaw et al., 2008).

In short, the JRF minimum income standard suggests that there are certain kinds of expenditures we should aim to protect as being essential to a 'decent life'; and others which may be considered unnecessary and hence could potentially be excluded from our consumption without compromising our ability to live well. This is potentially a very useful place from which to answer the question set out at the beginning of this chapter. What is the carbon footprint of a decent life? How much carbon would we need – at this point in time, in this kind of society – to provide everyone in the UK with a decent standard of living, taking into account both material and social needs.

In order to answer this question we used the Surrey Environmental Lifestyle Mapping framework developed in the RESOLVE project at the University of Surrey to estimate the greenhouse gases (GHGs) produced both in the UK and abroad during the production and distribution of all the goods and services purchased by UK households according to the minimum income standard defined by the JRF expenditure budgets. The results from this modelling exercise provide an indication of how much of the carbon that is used in our current consumer lifestyle is necessary for a 'decent' life, and how much may be thought of as 'discretionary' and could potentially be eliminated.

Of course, this exercise is purely hypothetical. It does not suggest either that everyone would agree on such a life, or that the components would be the same for everyone. The political viability of persuading people to forego 'discretionary' consumption is beyond the scope of this chapter; as is the challenge of providing the economics of a reduced consumption society. Nonetheless, the exercise is valuable in teasing out some of the difficult questions associated with achieving deep cuts in carbon emissions in response to climate change.

The organisation of this chapter is as follows. In the next section we describe the JRF budgets in more detail, setting out what is included and excluded in them, and comparing them to current expenditure levels. Next we briefly describe the methodology used to estimate the GHG emissions that would arise in the production and distribution of all the goods and services required for a decent life. Having done that we present a carbon footprint for the UK, calculated on the assumption that all UK households consume according to the expenditure baskets defined by the JRF 'decent life' and compare this against the UK Climate Change Act interim targets. Finally, we reflect on the implications of our analysis on the question of sustainable living, materialism and the 'good life'.

The income standard for a 'decent life'

The JRF study produced highly detailed weekly expenditure budgets for eleven types of households.[1] With respect to food, a nutritionist checked the menus to ensure they met current government guidelines for healthy eating. An example of the weekly meat allocation for a lone parent with one child (toddler) is 150g stewing steak; 400g beef mince; 62g bacon; 128g pork sausages; 175g chicken breasts; and 34g cooked chicken. All budgets contained some alcohol, the majority to be consumed in the home. For example a couple with one child is allocated 4 cans of Fosters' lager; 4 cans of Thwaites' Draught; and one bottle of Chilean white wine per week (presumably for the parents not the child).

A car was deemed unnecessary for the minimum standard of living to be met, and each budget includes a weekly bus pass for each household member (except for pensioners and small children who travel free). All budgets also included provision for taxi hire, to cover specific trips such as weekly supermarket visits, late night journeys, or emergency hospital visits. Each budget contained a one-week budget holiday in the UK.

Household fuel use in the budgets was calculated at the level necessary to 'maintain health and well being of the householders and the fabric of the home' (Oldfield, 2008). Table 4.1 shows the type of dwelling in which each family type is assumed to reside, and its prescribed heating regime. Each household type is assumed to occupy a dwelling that is closely matched to its family size. It should be noted that this is not currently the reality in the UK, where many households, even those on low incomes, and especially pensioners, have an extra bedroom above the number specified in Table 4.1.

The heating regimes are based on World Health Organisation guidelines. Average fuel use and costs were estimated using the Building Research Establishment Domestic Energy Model (BREDEM 12) (Anderson et al., 2002). It was assumed that cavity wall insulation, loft insulation

Table 4.1　Housing specifications and heating regimes

Family type	Dwelling and heating description
Single male, female	1-bed mid terrace ground floor flat, heating period 9 hours day, at 20°C.
Couple with no children	2-bed ground floor flat, heating 6 hours day, 12 at weekends, at 21°C.
Single pensioner	1-bed mid terrace ground floor flat, heating 21°C 16 hours a day.
Couple pensioner	2-bed ground floor flat, heating period 16 hours day, at 21°C.
Lone parent one (toddler) child Couple one (toddler) child	2-bed end terrace house, heating period 16 hours day, at 21°C.
Lone parent or couple two children, pre-school, primary school	3-bed mid terrace, heating period 16 hours day, at 21°C. (pre school child in family)
Lone parent or couple three children, pre-school, primary school, secondary school	4-bed house, heating period 16 hours day, at 21°C. (pre school child in family)
Couple four children, toddler, pre-school, primary school, secondary school	4-bed house, heating period 16 hours day, at 21°C. (pre school child in family)

Source: Oldfield (2008).

and double glazing were installed in each dwelling. Again, it should be pointed out that this differs from the reality today. This assumption, alongside the close fit of family size to number of bedrooms, resulted in the estimated household fuel consumption of each household type being below actual expenditure as shown in the Expenditure and Food Surveys for equivalent households (Oldfield, 2008).

To get a feeling for expenditures according to the JRF budgets, it is useful to compare them against the average actual levels of expenditures across household types in the UK at the moment. Figure 4.1 shows how the JRF budgets compare with expenditures reported in the Expenditure and Food Survey (EFS),[2] excluding housing and childcare. It can be seen, for example, that among single people of working age about two-thirds of families spend more according to the EFS than they do in the JRF budget. Similarly, among pensioner couples and couples with two children about three quarters spend more. However, among lone parents over half of families spend less at the moment than predicted by the JRF budget. This indicates that some family types are, in monetary terms, better off on average in the budget than they are in reality today (Bradshaw et al., 2008). Or equivalently that at present these households do not yet have the requirements for a decent life.

It should be noted that the minimum income standard budgets are for individual 'average' households in Britain. It is clearly acknowledged by the JRF authors that actual budgets would, of course, need to vary within this minimum standard. For example, many rural dwellings are not connected to the gas mains, and in rural locations transport may be less convenient and reliable, and also more expensive. Besides these circumstantial factors, individual choice over particular constituents in the basket of consumption could vary widely. Nonetheless, the minimum income standard is supposed to represent a level of consumption that could in principle deliver a 'decent life'.

The carbon footprint of a decent life

The carbon footprint of each household type was calculated by estimating the GHG emissions that arise both in the UK and abroad in the production and distribution of all the goods and services purchased according to the budget. This was done using the Environmental Input-Output sub-model within the Surrey Environmental Lifestyle MApping (SELMA) framework (Druckman and Jackson, 2010).

Once the GHG emissions for each type of household had been estimated, total emissions across the UK were estimated by summing up the

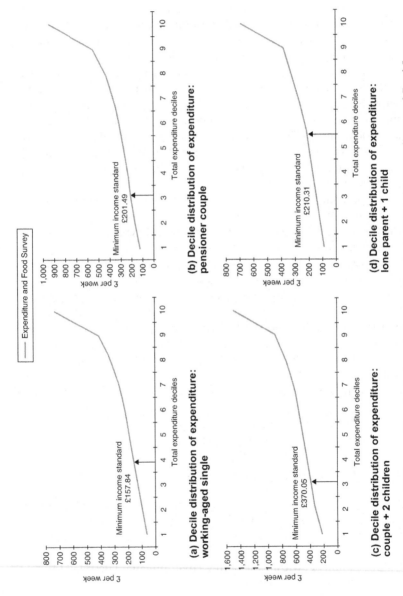

Figure 4.1 Comparison of JRF expenditure budgets against expenditure reported in the Expenditure and Food Survey
Source: Bradshaw et al. (2008).

emissions for each household type on the basis of the proportion of UK households represented by each type in the UK as shown in Table 4.2.[3] The results of this exercise are striking. Figure 4.2 suggests that the carbon footprint for a decent life is about 37 per cent lower than the current average carbon footprint of UK households. To be quite precise, the average household footprint under the minimum income standard is 17 tCO2-e as compared to actual (historical) consumption-based emissions of 26 tCO2-e.

To give some insight into the implications of the minimum income standard for carbon emissions it is instructive to examine which expenditure categories show major emissions reductions over the actual average UK emissions. Travel is a prominent change, since the minimum income standard assumes that the use of motor vehicles is not deemed necessary for a decent life. This means that emissions from personal transportation fuels are zero whereas in reality in 2004 the UK mean personal transportation emissions were around 2.6 tCO2-e, or ten per cent of the total carbon footprint. Furthermore, in the minimum income standard it is assumed that holidays would be taken in the UK, eliminating aviation emissions[5] which were responsible for around five per cent of UK household emissions in 2004 (Druckman and Jackson, 2009).

Household fuel use is another important category. Emissions from electricity, gas and other household fuels are reduced by 45 per cent in the expenditures defined by the minimum income standard. This is

Table 4.2 Distribution of minimum acceptable standard of living budget family types

	Family type	% of total UK households
A	Single pensioner	16.4
B	Couple pensioner	12.2
C	Single working age	16.0
D	Couple working age	18.4
E	Single parent, 1 child	3.7
F	Single parent, 2 children	2.8
G	Single parent, 3 children	0.9
H	Couple parents, 1 child	8.0
I	Couple parents, 2 children	11.2
J	Couple parents, 3 children	3.5
K	Couple parents, 4 children	0.9
	Other[4]	6.0
	Total	**100**

Source: Bradshaw et al. (2008). For more details see Druckman and Jackson (2010).

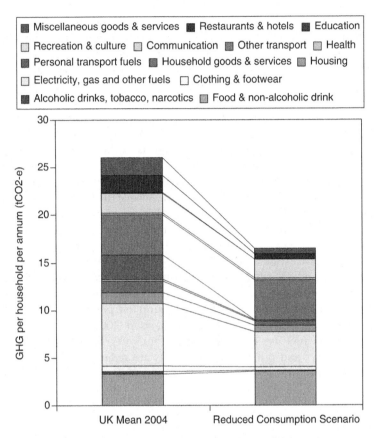

Figure 4.2 Comparison of mean GHG emissions per household for UK mean in 2004 against the minimum income standard emissions

achieved partly by assuming that a higher standard of thermal insulation of dwellings is achieved in the housing stock and partly because there is a closer fit of dwelling size to number of inhabitants, as described earlier. Other categories of expenditure that are reduced in the minimum income standard are restaurants and hotels (reduced by 69%); miscellaneous goods and services (68%); and household goods and services (60%). These categories are deemed by the JRF study to represent largely discretionary purchases that are not essential to a decent life.

Figure 4.2 illustrates that carbon emissions would be reduced in 12 out of the 14 categories of expenditure. Only one category significantly defies this trend: emissions due to food and non-alcoholic drink, which are estimated to *increase* under the assumption of a minimum income standard by around seven per cent. This finding reflects the fact that the specified diets, which were checked by nutritionists, including more fruit and vegetables than average 2004 UK diets.

It is worth comparing the emission reductions achieved here against the emission targets established by the UK Government. The UK has passed legislation to make legally binding GHG emissions reductions by 2050. The Act also established the Committee on Climate Change, which is an independent, expert body whose role is to advise Government on setting and meeting carbon budgets. The Committee has recommended that the UK should reduce emissions of all greenhouse gases by at least 33 per cent in 2020 relative to 1990 levels, and that this should be increased to 42 per cent relative to 1990 if a global deal to reduce emissions is achieved (CCC, 2008).

The reduced carbon footprint is 37 per cent lower than the actual 2004 footprint. As Figure 4.3 illustrates, this represents a third reduction in GHG emissions relative to 1990 emission levels. Taking into account the further reductions that could be achieved, for example by

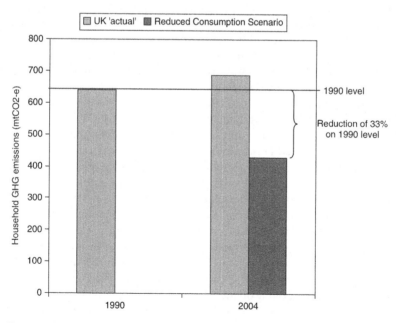

Figure 4.3 Minimum income standard GHG emissions compared against UK GHG reductions targets

technological shifts in the energy supply industry, it is clear that the UK target of at least a 34 per cent reduction of GHGs on 1990 levels by 2020 could be feasible within the household sector, without compromising the requirements for a decent life, as defined by the JRF study.

It should be noted here in passing, perhaps, that lower consumption in one nation or region does not in itself ensure that global emissions are reduced. Trade is global and UK household consumption must be considered as part of the global picture. Alcott (2008) has noted that lower consumption in one region may have the effect of lowering prices, which in turn stimulates demand by other regions. For effective global reduction in GHG emissions, reductions in the emissions of developed countries must be accompanied by globally agreed emissions quotas.

Materialism, sustainability and the decent life

The last 50 years have seen an unprecedented rise in consumerism. In 1957 the British prime minister, Harold Macmillan, declared that Britons had 'never had it so good'. He went on to say 'Go around the country, go to the industrial towns, go to the farms and you will see a state of prosperity such as we have never had in my lifetime – nor indeed in the history of this country'. Since that time real incomes have doubled, enabling a shift in expenditures from more mundane subsistence items such as food, clothing, household goods and housing, towards more discretionary expenditures such as recreation, entertainment and eating out in restaurants (Jackson and Papathanasopoulou, 2008).

Despite this material success, the percentage of UK citizens reporting themselves 'very happy' declined from 52 per cent in 1957 to 36 per cent before the recent recession[6] and rates of stress and depression are increasing (Naish 2008; Wilkinson and Pickett, 2009). Whereas the UK was ranked sixth in the world according to gross GDP, it was ranked only 108th using an index built from three different indicators (life expectancy, ecological footprint and subjective well-being) (Marks et al., 2006).

What emerges from these findings is the now much-advanced paradox: that the expansion of material consumption – and the materialistic aspirations that support it – do not appear to be making us any happier. Since the material consumption levels of the affluent West are also deeply inequitable and clearly unsustainable, there would seem to be something of a double dividend from consuming less (Jackson, 2005). Or at the very least, there is a clear premium on identifying visions of the good life in which human beings have an equal opportunity to flourish, within the finite ecological limits of the planet (Jackson, 2009).

One starting point for developing such a vision is to understand what we mean by a 'decent life' and determine the minimum material requirements for it. This chapter has set out one possible approach to that task. By identifying the basket of consumption goods deemed necessary not only to survive in physiological terms but to thrive in social terms, the JRF study provides the foundation for assessing the carbon footprint of a decent life. As we saw in the previous section, the GHG reductions achieved by moving to such a scenario are considerable.

It is perhaps worth exploring further how the ensuing lifestyles might differ from current lifestyles, particularly in relation to the role that material aspirations play in our lives. Firstly, of course it is clear that the material demands associated with leading a decent life are not exhausted by subsistence needs: food, clothing and shelter. On the contrary, many of the goods that we buy play important symbolic roles in our lives contributing to vital 'social conversations' that allow us to participate in the life of society (Douglas, 2006; Jackson, 2006). We use goods to mediate our relationships with friends and family; to attract sexual partners; to show allegiance to social groups; and to mark important occasions such as birthdays through, for example, the giving of gifts. Material goods are also implicated in the creation and maintenance of identity and the marking of status in society.

Although it is clear that status is important in every kind of society, and that material goods have often played some role in marking status, some have argued that modern consumer society is distinctive in playing out so many psychological needs for belonging, self esteem, respect, self-identity and social status through the 'language' of material goods (Jackson, 2006; Offer, 2006). But this never-ending quest also brings anxiety, much of which rests with the constant need to win, demonstrate and retain status in a fast moving world (de Botton, 2004). Furthermore, it has been shown that competition for status is higher in more unequal societies, as the struggle to 'keep up with the Joneses' becomes more acute when the gaps between the rich and poor widen (James, 2007; Wilkinson and Pickett, 2009). As Hirsch (1977) pointed out over 30 years ago, positional consumption is a net zero sum game (Naish, 2008). Although we may change our relative position by acquiring a bigger house, a better car or the latest consumer gadget, the status gains are at the expense of those around us, so net well-being does not change. Indeed it may even decline, since a loss of status is believed to have a bigger psychological impact on well-being than a gain in status.

There is moreover a dynamic element to this positional competition that pulls more and more material goods through the economic system

(Jackson, 2009). As consumer novelties move from being high-status luxuries to common necessities, those who strive to stay ahead of the curve must move on to the next new item, leading to a never ending pursuit of yet more novelty. This results in a great deal of resources being used and GHGs emitted, with no net gain in either individual status or social well-being.

In these circumstances, the potential gains from a more equitable and less materialistic distribution of goods and services, like the one defined in this chapter, should be apparent. It is still of course vitally important to ensure that basic subsistence needs are widely met, with great care and attention being given to, for example, the nutritional value of the prescribed food allocations, and adequate warmth in dwellings. But equally, it is clear that allowance is made for the goods needed to mark important occasions by buying gifts, the food and drink required for participation in social events, and the mobility needed to visit one's friends and family.

The JRF budgets are explicitly constructed to facilitate these needs. On the other hand, they are clearly not capable of meeting the expanding material aspirations of our present consumer society. Everything in the budgets was considered to be 'necessary' whereas in our current culture of consumption many goods are purchased that are subsequently found not to be needed (Trocchia and Janda, 2002). This phenomenon of 'over-consumption' is illustrated by the plethora of unused items advertised on websites such as Freecycle,[7] Ebay,[8] and the expansion of the personal storage industry (Naish, 2008). In the JRF budgets the vast quantities of purchases that result in goods being 'under-consumed' or even never used at all are eliminated. One of the benefits of this, of course, is that emissions of the GHGs that arise in the production and distribution of these goods are also eliminated.

On the hand, on the surface at least, the JRF budgets appear to exclude provision for goods which are seen mainly as positional goods or status markers. This does not necessarily mean of course that a society defined according to the minimum income standard for a decent life is entirely devoid of status hierarchies. For instance, it is clearly possible to conceive of alternative ways in which status may be marked, even when incomes are equalised and material distinctions are eroded. Although material goods are used widely for marking status in most cultures, Western society has recently taken this to extremes. For example, at the beginning of Margaret Thatcher's term as UK Prime Minister at the end of the 1970s, status was derived mainly through a person's job title, family history, and school attended. In contrast, by the end of her era some eleven years later, status was defined increasingly by material possessions (Bauman, 1998; Goodall, 2007).

It should be noted that the remit of the JRF study did not include the environmental aspects of consumption. Therefore the budgets were designed largely without regard to resultant environmental impact. Interestingly, however, many of the expenditures reduced or eliminated are those that have high-energy intensity.

For example, the study assumed insulation levels in dwellings above the current national average[9] and, accordingly, consumption of gas for central heating is assumed to be lower than current mean UK levels. The study also assumed a closeness of fit between family size and dwelling.[10] Conditions in the UK are slowly moving in these directions: recent statistics show a trend towards the construction of dwellings with a smaller number of bedrooms[11] and the UK government has a policy to improve the thermal efficiency of the housing stock (HMG, 2009). However, the assumptions regarding housing represent a radical step change from current conditions.

Perhaps more problematically, the study assumes that personal vehicle transport is not necessary, and instead provides for purchasing bicycles and bus passes, along with taxi hire once a week for journeys that could otherwise be difficult. It is highly contentious to assume that participation in society is possible without private vehicle use, particularly in a society structured as ours is today and heavily reliant on infrastructures of private transport. Some studies have estimated that private vehicle use could be reduced by around 20 per cent – even within current transport structures (Goodwin, 2008). But beyond this, without major infrastructure changes, which are slow and costly to implement, participation in society may be jeopardised. Adjustments could be made to take account of this, but for the purposes of this exercise we have stayed faithful to the JRF study assumptions.

Likewise, clear differences in lifestyle are implied in other specific sectors, particularly with regard to food. Our analysis estimates that 22 per cent of the total carbon footprint for a decent life would be attributed to food and non-alcoholic drink. A diet with less meat and dairy foods would result in lower emissions, as livestock have been shown to account for a significant proportion of GHG emissions (Garnett, 2009). This could be done without jeopardising nutritional standards. For example, in a study which compared meals with comparable nutritional values, a meal made from potatoes, carrots and dry peas was estimated to have nine times lower emissions than a meal with tomatoes, rice and pork (Carlsson-Kanyama, 1998).

It should be noted that the JRF study did not challenge the underlying current consumer culture of the 'throw-away' society (Cooper, 2005). For example, it is assumed that adults' socks last for just one year. Clearly there is considerable potential to extend product lifetimes through better design, greater durability and more effective skills and capabilities for repair and maintenance. Likewise it is clear that considerable investment would be needed to increase the energy efficiency of dwellings, improve public transport and change the planning and infrastructure of provision.

The exercise in this chapter is primarily a tool for thinking about the carbon footprint for a decent life. Hence it is not our intention to map out a definitive socio-technological transition pathway or detailed policy prescriptions for how any specific shift in consumption patterns might come about in the UK. Nevertheless, some consideration of the policy issues is a useful addition to the picture.

In the first instance, it is probably worth reiterating that a totally egalitarian 'decent life' is an obvious abstraction. A completely 'flat' society in which every household has a clearly defined basket of consumer goods has no credibility at all. All the same, this exercise could be used to define the baseline for an equitable 'carbon budget' at the household level. So long as each household has the ability to design a decent life within this carbon budget, such an exercise could provide a basis for fairness in the distribution of carbon within an overall cap. Individual choice – and indeed trade of carbon allowances – would be a perfectly reasonable design feature of such a system.

Nonetheless, it is clear that even this looser assumption of choice and tradeability is radically different from the current situation in the UK, which saw significant increases in inequalities during the 1980s which will be hard to reverse (Goodman and Oldfield, 2004). Policies to reduce inequality therefore emerge as an essential precursor to a shift in the direction indicated here. These could include revising income tax structures, improved access to good education and anti-discrimination measures. As highlighted by Wilkinson and Pickett (2009), the dividends that can be expected from a more equal society in terms of well-being are considerable.

As discussed above, the marking services of consumer goods are fundamental to society, but while much of this happens through material goods in our society, the carbon implications of this are unsustainable. This suggests a need to create alternatives to material markers. Status, for example, could be marked through less materialistic measures, such as creative or intellectual accomplishment, courage, or service to others.

Such a change relies on a significant shift in social norms but it is also worth remarking here that some evidence for shifts in consumption norms does exist: examples include the 'voluntary simplicity' movement (Huneke, 2005), the 'downshifting' movement (Hamilton, 2003) and 'intentional communities' such as Findhorn.[12] All these lifestyle movements have in common a reduced emphasis on consumerism with the aim of achieving well-being through less materialistic means. An indication of the extent to which social norms are changing were given by a survey in which a quarter of British adults aged 30–59 gave affirmative answers to the question 'In the last ten years have you voluntarily made a long-term change in your lifestyle, other than planned retirement, which has resulted in you earning less money?' (Hamilton, 2003: 12). When asked by how much their incomes fell as a result of the life-change, the average reduction was 40 per cent. This is a substantial decline, suggesting a major life change. Although starting from a high consumerist base, Hamilton's survey suggests that there is at least some foundation for a move towards less-materialistic lifestyles.

One key feature of the scenario depicted in Figure 4.2 is that overall household expenditure would be significantly lower than it is at the moment, unless the price of goods changed considerably. This in its turn implies reduced levels of industrial turnover. Full analysis of the macro-economics required to achieve this is inevitably beyond the scope of this chapter, although one of the authors has addressed this issue in depth elsewhere (Jackson, 2009). Suffice to say here that one way to move towards this might be through a general reduction in working hours, whether through a shorter working week, a stepping down to one income per household, or a combination of both (Coote et al., 2010; Hayden and Shandra, 2009). Shorter working hours could also mean more time spent caring for children and the elderly, and for recreation and leisure. This in turn, may be expected to reduce levels of stress and increase levels of well-being. In short a decent life might well turn out to be a happier life.

Conclusion

The underlying premise in this chapter is that remaining within ecological limits demands a serious consideration of the possibility of consumption restraint in the richer nations such as the UK. Our aim was to explore what GHG emissions would be required to support a 'decent life' under such constraints. For such lifestyles to become a reality we assumed that they must meet an acceptable standard with regards to providing subsistence needs and also the social need to participate in society. We have drawn on a report by the Joseph Rowntree Foundation that drew up detailed expenditure budgets for a minimum acceptable income standard in the UK. We used these budgets to estimate the GHGs that would be emitted (in the UK and elsewhere) in the production and distribution of all the goods and services that are purchased according to the budgets. Our calculations have estimated that under these assumptions GHGs would be around 37 per cent lower than mean household GHG emissions in 2004.

In material terms, the scenario sketched in this chapter is of a hypothetical, egalitarian society, which provides the basics for people to live well and participate in society. Much of the 'over-consumption' that we associate with consumer lifestyles, including the relentless pursuit of status through displays of material goods, would be reduced. At the same time, the intention is to protect the requirements for a decent life, and there are some indications that, with less competition and fewer material burdens, well-being in such a society may even increase.

Public acceptance of reduced consumption would depend critically on shifting norms. This is an ambitious task. But evidence exists in a growing interest (if from a small base) in moving towards a less consumer-focused way of life. This offers a promising indication that less materialistic lives could become acceptable, provided appropriate alternatives were offered through which people could flourish as human beings.

Significant structural changes would be essential to achieve these outcomes: changes that are nonetheless very much in line with those required to move to a low-carbon economy. Specifically, substantial economic investment is required to improve public transport and the thermal efficiency of the UK housing stock, to such an extent that could be termed 'visionary' compared to current UK investment plans.

In summary, the conclusion from this study is optimistic. A shift towards a society less focused on status-driven consumerism is essential. Such a society would have to prioritise the task of providing capabilities for flourishing in less materialistic ways (Jackson, 2009). In particular it would need to build strong avenues for social participation and to renew a sense of common citizenship and purpose (Sandel, 2009). This task requires strong leadership and significant investment. But if supported by specific structural changes, our analysis suggests that significant reductions in GHG emissions could be achieved without jeopardising social well-being.

Notes

We are grateful to Dr Glen Peters for his kind provision of amendments to the GTAP GHG emissions data for selected countries. The research is supported by funding from the ESRC Research Group on Lifestyles Values and Environment (RESOLVE) (Grant Number RES-152-25-1004).

1. Each budget contains hundreds of costed items and allowances for activities and services purchased by each household type. 90 per cent of items were costed at national chain stores such as Tesco and Argos. For non-food items the lifetime of each object was estimated (taking account of its quality and type of usage) and its cost was spread over its estimated lifetime. The budgets provide generous allowances for childcare (for example, £135.05 per week for a lone parent with one child).

2. In order to undertake this comparison Bradshaw et al. (2008) merged five years of data from the Expenditure and Food Survey (EFS) (2001/2 – 2005/6) with each year's data being uprated to April 2008 using the commodity price index.

3. It was assumed that N. Ireland has same distribution of types of households as Britain.

4. 'Other' was ignored and the percentage of household types A–K re-scaled accordingly.

5. Aviation is included in 'Other transport' in this study.

6. See Ruut Veenhoven's 'World Happiness Database' available on the web at www2.eur.nl/fsw/research/happiness.

7. Freecycle is a grassroots and entirely non-profit movement of people who are giving (and getting) 'stuff' for free in their own towns. Its aim is to encourage re-use and to reduce the quantity of goods sent to landfill. See http://www.freecycle.org/.

8. See http://www.ebay.co.uk. A recent study has found that the percentage of unused products offered at eBay ranges from over 85 per cent of books to around 12 per cent of ICT and Consumer Electronics (Erdmann and Henseling, 2009).

9. The heating regime was based on the requirements in social housing which has on average higher thermal ratings than the private sector (Oldfield, 2008).

10. Of course, smaller family sizes result in the loss of the economies of scale that are generally achieved by larger households (Druckman and Jackson, 2008). A more detailed discussion of this is beyond the remit of this chapter.

11. Housebuilding completions by number of bedrooms, England. Office for National Statistics, available online at http://www.statistics.gov.uk/cci/nugget.asp?id=1768 (accessed 14 January 2010).

12. See http://www.ecovillagefindhorn.com/.

References

Alcott, B. (2008) 'The Sufficiency Strategy: Would Rich-World Frugality Lower Environmental Impact?', *Ecological Economics*, 64(4): 770–86.

Anderson, B., P. Chapman, N. Cutland, C. Dickson, S. Doran, J. Henderson, P. Iles, L. Kosima and L Shorrock (2002) *BREDEM-8 Model Description: 2001 Update*, Garston, Watford: BRE.

Bauman, Z. (1998) *Work, Consumerism and the New Poor*, Milton Keynes: Open University Press.

Bradshaw, J., S. Middleton, A. Davis, N. Oldfield, N. Smith, L. Cusworth and J. Williams (2008) *A Minimum Income Standard for Britain: What People Think*, York: Joseph Rowntree Foundation.

Carlsson-Kanyama, A. (1998) 'Climate Change and Dietary Choices – how can Emissions of Greenhouse Gases from Food Consumption be Reduced?', *Food Policy*, 23(3–4): 277–93.

CCC (2008) *Building a Low-Carbon Economy – The UK's Contribution to Tackling Climate Change*, London: Committee on Climate Change.

Cooper, T. (2005) 'Slower Consumption: Reflections on Product Life Spans and the "Throwaway Society"', *Journal of Industrial Ecology*, 9(1): 51–68.

Coote, A., J. Franklin and A. Simms (2010) *21 Hours: Why a Shorter Working Week Can Help us All to Flourish in the 21st Century*, London: New Economics Foundation.

De Botton, A. (2004) *Status Anxiety*, London: Hamish Hamilton – Penguin Books.

Douglas, M. (2006) 'Relative Poverty, Relative Communication', in A. Halsey (ed.) *Traditions of Social Policy*, Oxford: Basil Blackwell.

Druckman, A. and T. Jackson (2008) 'Household Energy Consumption in the UK: A Highly Geographically and Socio-Economically Disaggregated Model', *Energy Policy*, 36(8): 3167–82.

Druckman, A. and T. Jackson (2009) 'The Carbon Footprint of UK Households 1990–2004: A Socio-Economically Disaggregated, Quasi-Multiregional Input-Output Model', *Ecological Economics*, 68 (7): 2066–77.

Druckman, A. and T. Jackson (2010) 'The Bare Necessities: How Much Household Carbon do we Really Need?', *Ecological Economics*, 69(9): 1794–804.

Erdmann, L. and C. Henseling (2009) 'From Consumer to Prosumer – Development of New Trading and Auction Cultures to Promote Sustainable Consumption', 5th International Conference on Industrial Ecology, 2009 ISIE Conference: Transitions Toward Sustainability. Lisbon, Portugal.

Garnett, T. (2009) 'Livestock-Related Greenhouse Gas Emissions: Impacts and Options for Policy Makers,' *Environmental Science & Policy*, 12(4): 491–503.

Goodall, C. (2007) *How to Live a Low-Carbon Life: The Individual's Guide to Stopping Climate Change*, London: Earthscan.

Goodman, A. and Z. Oldfield (2004) *Permanent Differences? Income and Expenditure Inequality in the 1990s and 2000s*, London: The Institute for Fiscal Studies.

Goodwin, P. (2008) 'Policy Incentives to Change Behaviour in Passenger Transport', OECD International Transport Forum, Leipzig, Germany, Transport and Energy: The Challenge of Climate Change, The Centre for Transport & Society, University of the West of England, Bristol, UK.

Hamilton, C. (2003) 'Downshifting in Britain: A Sea-Change in the Pursuit of Happiness', Manuka, Australia, The Australia Institute, Discussion Paper Number 58.

Hayden, A. and J. Shandra (2009) 'Hours of Work and the Ecological Footprint of Nations: An Exploratory Analysis,' *Local Environment*, 14: 575–600.

Hirsch, F. (1977) *Social Limits to Growth*. Revised edition (1995), London and New York: Routledge.

HMG (2009) *The UK Low Carbon Transition Plan: National Strategy for Climate and Energy*, London: TSO.

Huneke, M. (2005) 'The Face of the Un-Consumer: An Empirical Examination of the Practice of Voluntary Simplicity in the United States', *Psychology & Marketing*, 22: 527–50.

Jackson, T. (2005) 'Live Better by Consuming Less? Is There a "Double Dividend" in Sustainable Consumption?', *Journal of Industrial Ecology*, 9(1): 19–36.

Jackson, T. (2006) *Earthscan Reader in Sustainable Consumption*, London: Earthscan.

Jackson, T. and E. Papathanasopoulou (2008) 'Luxury or "Lock-in"? An Explanation of Unsustainable Consumption in the UK: 1968 to 2000', *Ecological Economics*, 68(1–2): 80–95.

Jackson, T. (2009) *Prosperity Without Growth: Economics for a Finite Planet*, London: Earthscan.

James, O. (2007) *Affluenza*, London: Vermilion.

Marks, N., A. Simms, S. Thompson and S. Abdallah (2006) *The Un-Happy Planet Index: An Index of Human Well-Being and Environmental Impact*, London, New Economics Foundation and Friends of the Earth.

Naish, J. (2008) *Enough: Breaking Free from the World of More*, London: Hodder & Stoughton.

Offer, A. (2006) *The Challenge of Affluence*, Oxford: Oxford University Press.

Oldfield, N. (2008) *The Fuel Budget Standard*, York, Joseph Rowntree Foundation.

Sandel, M. (2009) *A New Citizenship: The Reith Lectures 2009*, London, BBC.

Townsend, P. (1979) *Poverty in the United Kingdom: A Survey of Household Resources and Standards of Living*, Harmondsworth: Penguin Books Ltd.

Trocchia, P. and S. Janda (2002) 'An Investigation of Product Purchase and Subsequent Non-Consumption', *Journal of Consumer Marketing*, 19(3): 188–204.

Wilkinson, R. and K. Pickett (2009) *The Spirit Level: Why More Equal Societies Almost Always Do Better*, London: Allen Lane – Penguin Group.

5

Transport and Mobility Choices in 2050

Stephen Potter

Transport is possibly the most problematic area with regard to achieving a low-carbon society. It is the UK's fastest-growing source of CO_2 emissions, with domestic transport contributing 27.5 per cent of the UK CO_2 emissions (DfT, 2007, Table 3.8). The 2006 UK Stern report (Stern, 2006) noted that between 1990 and 2002, transport was the fastest growing source of carbon emissions in OECD countries (a growth of 25%) and the second fastest growing sector in non-OECD countries (36% growth). Rather than declining over the next 40 years, the trend is for transport CO_2 emissions to grow, particularly in non-OECD countries, where their share of global emissions is anticipated to grow from one-third to one-half by 2030.

The transport sustainability challenge

Over 90 per cent of the UK's transport CO_2 emissions come from road transport (Table 5.1). Passenger cars remain the biggest source of CO_2, but road freight emissions are significant and have risen by 23 per cent over the last ten years compared with roughly static emissions from passenger cars. Rail produces less than two per cent of transport's CO_2 emissions, despite recent substantial rises in passenger-kilometres and freight carried.

Carbon emissions from aviation have also grown rapidly in the last decade, up by 60 per cent. If international aviation is included, it accounts for nearly a quarter of all transport's CO_2 emissions. The 2004 *Transport Policy White Paper* (DfT, 2004) noted that because emissions at altitude have a greater global warming effect, these now represent 11 per cent of the UK's total climate change impact. At currently predicted growth rates the aviation sector will constitute about a third of total UK climate change impact by 2050. So, even if all other sectors meet

Table 5.1   UK CO$_2$ emissions by source

Source	2005 Million tonnes of CO$_2$	Percentage of all UK domestic transport emissions	Percentage of all UK domestic emissions
Passenger Cars	69.9	54.2	12.6
Light duty vehicles (vans etc)	16.8	13.0	3.0
Buses	3.6	2.8	0.6
Heavy Goods Vehicles	28.6	22.2	5.2
Mopeds and motorcycles	0.4	0.3	0.1
Other road transport	0.6	0.5	0.1
All Road	(119.9)	(92.9)	(21.6)
Railways	2.0	1.6	0.4
Domestic aviation	2.5	1.9	0.4
Domestic shipping/ navigation	4.2	3.3	0.8
Total (Domestic)	129.0	100.0	100.0
International aviation	35.0		
International shipping	1.4		
Total	**165.4**		

Source: Department for Transport, 2007 Table 3.8.

government CO$_2$ reduction targets, air travel is a key environmental issue for the twenty-first century (Bishop and Grayling, 2003).

Given this situation, how might we achieve a low-carbon transport future? A list of greener transport initiatives is not difficult to compile. 'Low-carbon' vehicles and fuels are particularly attracting attention. The comprehensive E4tech report for the UK Department for Transport (E4tech, 2006), reviewed a range of vehicle technologies (including battery-electric, hybrid-electric and fuel cells) and a range of related fuels (gasoline, diesel, bioethanol, biodiesel and hydrogen). This study concluded that, compared to conventional petrol and diesel-engined cars, hybrid cars can cut carbon emissions by around 20 per cent. The use of low-carbon fuels offers greater improvements; bioethanol can cut CO$_2$ emissions per vehicle kilometre by 25 per cent, biodiesel by 45 per cent and hydrogen by 40 per cent or more. However, these improvements very much depend on the production methods used. But, crucially, almost all these technologies appear to fall short of the 60 per cent (or more) cut in all transport's CO$_2$ emissions that would be needed to achieve a 2050 low-carbon society.

International studies (e.g. EUCAR et al., 2005) have produced similar results. In 2002, the UK Government has set a target that low-carbon cars should represent ten per cent of all car sales by 2012 (DfT, 2002) and in 2005 announced the Renewable Transport Fuels Obligation, requiring suppliers to source five per cent of transport fuel sales from renewable sources by 2010/11. This formed a major part of transport's contribution to the 2006 Energy Review (DTI, 2006). It is notable that the commitment to low-carbon battery electric cars was the one transport policy measure to feature in the 2010 Conservative/Liberal Democrat Coalition pact.

An alternative approach to low-carbon fuels is to use fuel more efficiently. This is actually how hybrid cars cut carbon emissions, but there is a greater potential than the 20 per cent improvements they achieve. Well over a decade ago, Wemyss (1996) in his technological review considered that advances in vehicle technologies should allow cars to achieve 1.9 litres/100 km (150 mpg) within ten years. Yet, despite some progress in the fuel economy of new vehicles, there are still no cars on the market that achieve anywhere near this technically possible performance.

A separate range of measures involve more of a systems approach, advocating modal shift from cars to more energy efficient forms of transport including light rail and innovative public transport systems, public shared bicycle schemes, car pooling, car clubs and telecommuting. This is often coupled with proposals for planning controls to produce settlement patterns and conditions that will favour sustainable modes and disadvantage car use. In his comprehensive review of this and other approaches, Banister (2005, Chapter 6) cites case studies of cities that have achieved a ten per cent cut in car use through approaches utilising planning controls and public transport development.

A further systems approach is that of pricing mechanisms. Economists have long argued that the core problem is the under-pricing of the environmental costs of road and air transport and that marginal cost pricing should be adopted (Maddison et al., 1996; ECMT, 1997; Glaister and Graham, 2003). The arguments and evidence for a tax neutral programme of green fiscal reform was presented by the 2009 report of the UK Green Fiscal Commission, including a detailed briefing paper on transport taxation (Green Fiscal Commission, 2009).

Each of the above approaches to vehicle technologies, modal split, planning and fiscal reform could represent part of moving towards a 2050 vision of a low-carbon transport, but what might be the relative roles of each of these components? This chapter takes surface transport (which covers a major part of UK transport's carbon emissions) and undertakes a macro-level analysis exploring what sort of strategic approaches could deliver a genuinely low-carbon transport future. The intention is to set a specification and provide a framework to identify targets and strategies for development and deployment. In doing this, although technical and policy viability are necessarily taken into account, they are not the chapter's main focus. This is a 'backcasting' exercise starting from a definition of a 2050 low-carbon personal transport sector, and then exploring if various combinations of transport technologies and changes in travel can take us from our current position to one of sustainability. The process uses a simple equation model. This is purposely simple in order to provide a tool to develop understanding by anyone wanting to explore transport's sustainability challenges. This tool has been used in Open University environment courses (Potter and Warren, 2006 and Potter, 2007) and in stakeholder meetings to evaluate transport policy development.

Exploring the issue

At the moment, cars in the UK car fleet averages 9.1 litres/100 km with the best performing cars returning under 6 litres/100 km. Technically, the application of best current practice could improve fuel economy to an average of around 6 litres/100 km with the best being 4 litres/100 km. However, in practice, despite such designs being available for the last 15 years, only marginal improvements have arisen. The simple fact is that new cars are sold on their power, top speed, acceleration, style and equipment. An industry regime has emerged around these design features which uses improvements in fuel efficiency mainly to enhance performance rather than cut fuel consumption.

Alternative fuelled vehicles originally emerged in response to air quality concerns, particularly in the USA. These include designs for vehicles powered by electricity, compressed natural gas (CNG), liquid petroleum gas and hybrids, which combine electric and internal combustion drives. Although these fuels offer significant reductions in the emission of local air pollutants, in terms of carbon it is a mixed picture. As was noted above, E4tech (2006), estimated such technologies to produce between 20 per cent and 40 per cent less CO_2 compared to conventional petrol and diesel-engined cars.

In the long term, for some time virtually all commentators consider the use of hydrogen fuel cells linked to renewable energy generation as the ultimate ideal clean traction method for transport. Lane (2004) reviews the actual and anticipated performance of fuel cell cars and notes that using natural gas as a feedstock for a fuel cell would produce between 12 and 43 per cent less emissions compared to using natural gas in an internal combustion engine – itself cleaner than the best petrol or diesel technologies. However, fuel cell efficiencies vary considerably and improvement in emissions may be less than is widely claimed. In addition building a supply infrastructure for hydrogen vehicles would be a massive task (Berridge, 2009).

Even if more efficient vehicles were built and cleaner fuels used, would this be enough to achieve carbon sustainability by 2050? Cleaner technologies may be emerging, but whether they have sufficient scope is another question. One method is explore combinations of measures to cut environmental impacts is a simple environmental impact formula. Such a formula was originally developed by Paul and Anne Ehrlich (1990) and refined by Ekins et al. (1992), who uses the equation that environmental impacts are the product of:

$$\text{Population (P)} \times \text{Consumption (C)} \times \text{Technology (T)}$$

Environmental impacts are the sum of the number of people, how much each person consumes and the technology used in the goods and services they consume. This simple equation assumes that P, C and T are independent variables. In practice they do exert some influence upon each other; for example, increased wealth, expressed as consumption (C), can and does influence population growth; equally improvements to technology that reduce cost will stimulate more consumption (sometimes called the 'rebound effect' – see Herring and Roy, 2007). However, this criticism, although statistically valid, misses the point of such an approach, which is to explore at a very basic level, the implications of alternative scenarios in these three key variables. For example, at global level, it could be assumed that population grows by a third by 2050 and eventually stabilise at twice its current level. Global consumption might also double in this period. If we take the current situation as an index (i.e. everything is currently 1.0) and assume that technologies do not get any environmentally cleaner this results in:

Equation 1:

Current position:	P $\times$ C $\times$ T	=	Impacts
	1 $\times$ 1 $\times$ 1	=	1
In 50 years:	1.3 $\times$ 2 $\times$ 1	=	2.6

On this basis, environmental impacts will rise to 2.6 times current levels, so simply in order to stop them from worsening, the figure for 'Technology' has to drop to an index of 0.38. This simple little bit of mathematics suggests that the CO_2 impact of all goods and services need to be cut over the next 40 years to around a third of current levels simply to stop global environmental impacts getting worse. Indeed, if we were to allow for the rebound effect of improved technology further increasing consumption, then the Technology index would need to drop even further.

This environmental impacts formula approach has been developed by the author (Potter and Warren, 2006) and others (including Kwon and Preston, 2005) to analyse transport's carbon dioxide emissions. In this adaptation, 'Consumption' becomes a function of the number of car journeys per person and journey length. Technology can be expressed in terms of the CO_2 emissions produced per vehicle kilometre (which is a combination of the fuel economy of the vehicle and the carbon content of the fuel used). Thus the environmental impacts from motorised vehicles would be:

$$\text{Population} \times \text{Car journeys per person} \times \text{Length} \times \text{Emissions per vehicle km}$$

The baseline emissions situation expressed as an index would be:

Equation 2:

Population × Car Journeys × Length × Emissions per vehicle km = Total Pollution
$$1 \times 1 \times 1 \times 1 = 1$$

Again, this approach assumes independence between the variables, which is not entirely so. Improved fuel consumption (part of 'emissions per vehicle km') would reduce cost and therefore lead to some increase in the number of car journeys and their length. However this is a relatively minor impact on overall costs and, for the level of detail explored, the independence of variables is a reasonable simplifying assumption.

From this a 'business as usual' scenario can be developed. Here, a 20-year scenario is used rather than 40 years to 2050. The shorter timescale is used as it allows us to explore technologies and policies that can be more realistically assessed. However this work will use a CO_2 reduction target that represents a convergence path to a 60–80 per cent cut by 2050. It thus relates to a path leading to the full 2050 target.

Table 5.2   Indices of transport trends

	Index in 20 years
About a 5 per cent increase in population	1.05
Car journeys average about 630 per year (currently rising by 6–7 per year)	1.2
Journey length averages 11.1 km (rising at 0.15 km a year)	1.3
Fuel use averages 9.1 litres per 100 km across the car fleet (improving by 0.2 per cent a year assumed to improve to 8 litres per 100 km)	0.88

Source: Noble and Potter (1998) and Department for Transport (2007).

The current situation and trends for the UK car fleet are shown in Table 5.2. For fuel economy, a greater improvement than historically achieved has been assumed reflecting tightening EU standards and as increasing oil prices take effect.

The 'business as usual' scenario for 20 years time would result in the equation becoming:

Equation 3:

Population $\times$ Journeys $\times$ Length $\times$ CO_2 per vehicle km = Total pollution
$1.05 \times 1.2 \times 1.3 \times 0.88 = 1.44$

So, despite an improvement in fuel economy, CO_2 emissions increase by 44 per cent over the next 20 years.

Technical fix scenario

If there were a purely technical approach, affecting only the last part of the equation, then simply to stop emissions getting worse would require reducing the index for CO_2 per vehicle km to be 0.61. If current fuels are used this would mean improving average on road fuel consumption from the present 9.1 to 5.5 litres per 100 km. Even allowing for actual on-road fuel economy being some 20 per cent poorer than test figures, present high fuel economy designs could hit this target. However, such vehicles would have to be in very widespread use.

But such an improvement in vehicle fuel consumption would do no more than stop CO_2 emissions getting worse. If we are looking to a 60–80 per cent reduction in UK CO_2 emissions by 2050, then we need at least a 40 per cent cut in the next 20 years (which fits into the trajectory of the EU 2010 Copenhagen target for a 20% cut in ten years time on a 1990 emissions baseline).

A sustainability target for this formula model that is consistent with a 2050 low-carbon society would be for a 20-year target of a reduction of 40 per cent in transport's 1990 CO_2 level. In Britain, CO_2 from transport has already risen by ten per cent since 1990, so the index target needs to be adjusted down to 0.36 rather than 0.40. If we are to hit this target by using technical measures alone, it is a simple matter of working through the equation. The mathematics are that, if the travel growth parts of the formula were not altered, the 'CO_2 per vehicle km' index figure would need to be 0.22 in order to hit an overall index target of 0.36 for the personal transport system as a whole.

If this were to be achieved by fuel economy alone, then the UK car fleet would need to achieve an average fuel economy of two litres per 100 km. This is a very ambitious technical target and is getting close to the best claimed for small, lightweight hybrid-engined cars. Some cars might achieve such fuel economy but getting the entire car fleet to average this figure is another matter. An alternative would be to combine fuel economy with the development of low-carbon fuels. So if fuel economy could improve to around a fleet average of five litres/100 km and low-carbon fuels introduced that cut the carbon intensity of road fuels by 60 per cent, this would hit the target. Equation 4 shows this result, which splits out CO_2 per vehicle km into fuel use/km and Carbon Intensity/km.

Equation 4:

Population × Journeys × Length × Fuel use/vehicle km × Carbon Intensity = Total
1.05 × 1.2 × 1.3 × 0.55 × 0.4 = 0.36

Overall, this exercise suggests that, on their own, it is hard to envisage that either ultra-fuel efficient cars or the widespread adoption of low-carbon and renewable fuels could in 20 years deliver a sufficient improvement in CO_2 emissions. In particular, the current emphasis on low-carbon fuels represents a danger of just replacing petrol of diesel gas guzzlers with electric or hydrogen guzzlers. That will not get us on track for transport sustainability. A shift to lower carbon fuels needs to be combined with substantial improvements in fuel economy. But even if this were to be the case, the pace of technological improvement would need to rapidly increase and looks beyond that envisaged as achievable in this timescale. This is not to decry technological improvements. It is just that on their own they cannot realistically hit the sustainability target, even though they do offer a substantial reduction in CO_2 compared with current vehicles.

Modal shift scenario

If it looks like fuel and vehicle technology would require unrealistic improvements to deliver sustainability, what about moving to another part of the CO_2 – generation equation and explore the possible behavioural change policies, such as shifting a substantial amount of travel to public transport – a much advocated response to transport's environmental impacts. This could be through a variety of investment, planning, land use and fiscal measures.

With modal shift, a key question is by how much energy use and CO_2 emissions are reduced when people travel by transport modes other than the private car. For cycling, emissions are effectively zero, but for public transport, emissions depend strongly on the occupancy of trains and buses, which varies considerably. Figures are often quoted in terms of the seats in vehicles. Table 5.3 shows that in terms of seat kilometres, bus and rail have a 1.5–4 times improvement in energy efficiency over cars. However, this drops to only a zero to two times improvement when current occupancy is taken into account. Furthermore, the relative performance of car and public transport varies with journey purpose. For example, for commuting car occupancy is low and public transport vehicle occupancy very high, resulting in the average car commuting trip using over five times the energy of public transport (Potter, 2004). However for all trip purposes, the gap is narrower. A realistic assumption is that a transfer to public transport would increase vehicle occupancy and hence improve fuel efficiency per passenger kilometre, but overall more than a three fold improvement compared to car is unlikely.

At the moment in Britain, according to the National Travel Survey (Department for Transport, 2008), car use accounts for 64 per cent of trips, walking 25 per cent, bicycle one per cent, bus six per cent and train two per cent (with motorbike, air, taxi and other minor modes making up the balance). However, our interest is in motorised trips (which produce carbon dioxide); for these the car has an 88 per cent share, with bus at ten per cent and train at two per cent.

The environmental impacts equation can be adapted to proportionately cover all motorised passenger travel. Allowing for the relative

Table 5.3 Current energy use of transport modes (megajoules)

Mode	MJ per seat km	MJ per passenger km (average occupancy in UK)
Small Petrol Car	0.6	1.4
Medium Petrol Car	1.0	2.2
Large Petrol Car	1.3	2.9
Motorcycle	0.9	1.6
Bus	0.3	1.4
Rail	0.4	1.4

Source: Potter, 2004 and Potter and Warren, 2006.

energy efficiencies of the car, bus and train, Britain's baseline situation would be:

Equation 5:

Population × Journeys × Length × Emissions/vehicle km × modal share = Total

Car: $1 \times 1 \times 1 \times 1.1 \times 0.88$

Bus: $1 \times 1 \times 1 \times 0.3 \times 0.10 = 1$

Train: $1 \times 1 \times 1 \times 0.4 \times 0.02$

If, as before, the first stage in exploring this policy option is to assume an historical improvement in energy efficiency (to 88 per cent of current fuel used), together with policies to produce a modal shift resulting in the share for bus rising to 25 per cent and train to ten per cent with car is dropping to 65 per cent. These are the targets adopted in some planning studies and suggested by the Royal Commission on Environmental Pollution (RCEP, 1994). This is shown in equation 6

Equation 6:

Population × Journeys × Length × Emissions/vehicle km × modal share = Total

Car: $1.05 \times 1.2 \times 1.3 \times (1.1 \times 0.88) \times 0.65$ (car)

Bus: $1.05 \times 1.2 \times 1.3 \times (0.3 \times 0.88) \times 0.25$ (bus) $= 1.2$

Train: $1.05 \times 1.2 \times 1.3 \times (0.4 \times 0.88) \times 0.10$ (train)

So, even with this large modal shift, CO_2 emissions rise by 20 per cent. This may be a substantial improvement from the 50 per cent rise under the 'business as usual' scenario, but it is still an increase by a fifth. It would require an heroic change in travel behaviour to stabilise, let alone cut CO_2 emissions. So, just the same way that, on their own, vehicle technical improvements will fail to provide a sufficient environmental improvement, so also would the sole use of modal shift policies.

A fusion solution

If on their own neither behavioural changes nor technical improvements can viably hit the sustainability target, then what about fusing together the two? Equation 7 includes the modal shift figures and a large, but technically possible, improvement in the fuel emissions of all modes (through a combination of efficiency and switch to low-carbon fuels). This is represented in the fourth column of Equation 7. For cars a reduction to 30 per cent of current CO_2 emissions per vehicle kilometre is assumed. Although energy efficiency technologies would also benefit public transport as well as cars, there are some constraints. For example, light-weighting is less viable especially in rail vehicles, but switching to low-carbon fuels in fleets is possibly more immediately viable than for cars. Overall the model assumes improvements are slightly less for buses than for cars, and rail slightly less than this.

Equation 7:

Population $\times$ Journeys $\times$ Length $\times$ Emissions/vehicle km $\times$ modal share = Total

Car:	$1.05 \times 1.2 \times 1.3 \times (1.1 \times 0.3) \times 0.65$ (car)
Bus:	$1.05 \times 1.2 \times 1.3 \times (0.3 \times 0.4) \times 0.25$ (bus) = 0.43
Train:	$1.05 \times 1.2 \times 1.3 \times (0.4 \times 0.5) \times 0.10$ (train)

The result is getting close to the sustainability goal. This has an important policy lesson; whereas separately neither technical measures (low-carbon and fuel-efficient technologies) nor behavioural (modal shift) can provide an adequate improvement in CO_2 emissions, the combined effect is powerful. However, despite managing to get close to the sustainability goal, this still assumes very substantial technical and fuel switch improvements as well as a substantial model shift. Both could be eased if changes were made to yet other parts of the equation. These are parts that policy has tended to sideline. In particular, policies to promote public transport rarely consider journey numbers and length. It seems to be assumed that any increase in public transport use must be beneficial. This is not necessarily so. If an increase in public transport simply adds to transport intensity, then this is environmentally damaging.

A scenario can be explored assuming:

- Trip lengthening is halted for bus and car journeys.
- Rail trip length rises, to substitute for longer car journeys.
- A small rise in the number of motorised journeys.
- A less ambitious fuel economy improvement for cars and buses.

These final adjustments result in hitting the sustainability target in a more robust way as shown in Equation 8:

Equation 8:

Population × Journeys × Length × Emissions/vehicle km × modal share = Total

Car: $1.05 \times 1.1 \times 1.0 \times (1.1 \times 0.35) \times 0.65$ (car)

Bus: $1.05 \times 1.1 \times 1.0 \times (0.3 \times 0.45) \times 0.25$ (bus) $= 0.35$

Train: $1.05 \times 1.1 \times 1.2 \times (0.4 \times 0.5) \times 0.10$ (train)

Overall, in developing anything like a viable approach to hitting transport's sustainability goal, there needs to be a fusion of the role of increased fuel efficiency, modal shift and trip length reduction. This is consistent with the conclusions of an OECD/G8 study (EPA, 1998) exploring the reduction of all environmentally damaging emissions from transport to sustainable levels by 2030. This study concluded that a third of the reduction could be achieved by technical measures and two-thirds by demand management.

How quickly such a target might be achieved remains a further question. The above changes clearly cannot be achieved within a short timescale. With strong political will and social acceptance, it might be possible in 20 years, but a 30-year or more timescale seems more likely.

Conclusions

This exercise shows that, to fulfil the needs of sustainability, technical measures in isolation are likely to be ineffective. Equally even substantial modal shift to public transport cannot attain the sustainability target and will also be politically and socially hard to achieve. Trip length in particular is neglected as a focus for demand management measures.

A combined strategy, seeking to optimise technical improvements with demand management addressing trip length, trip generation and modal share can deliver the necessary improvement. This is in a realistic, though still tough, package. A real danger is that it may be politically easier to develop some technical measures (e.g. fuel switch) more readily than demand management. There would be a danger of the success of technical measures resulting in the neglect or abandoning of demand management policies. This back-casting exercise shows that if everything depends on one group of measures, then sustainable transport become unattainable even if technical improvements are pushed to unrealistic extremes. The foundations of longer-term and more tricky measures need to be put into place, while 'quick wins' are being implemented.

Transport policies at the local, national and international level need to blend technical improvements to vehicles with modal shift and also reduce the growth in journey lengths. Transport's environmental challenge is of such a magnitude that, unless substantial progress is made on all these fronts, we will inevitably fail to get on track for transport sustainability. This approach also highlights a fundamental weakness in the approach of transport policy, which assumes only slight adaptations in systems and behaviours. Measures to cut travel through increasing command and control regulation and the manipulation of people and societies run counter to our modern society. Regulation, fiscal and planning measures must play a part, but cannot be carried through to extremes. In terms of policy responses a 'fusion' approach seems also to be necessary. Traditional transport and planning policy measures need to be combined with other initiatives. Indeed, the key to transport sustainability may lie in finding alliances with social and economic trends towards the information society leading to the reinvention of how access is achieved. This has major implications for the nature of transport planning, which needs to shift towards a focus on service development, delivery and social marking. Indeed very concept of transport planning may cease to have much meaning in a low-carbon society.

References

Banister, D. (2005) *Unsustainable Transport*, London: Taylor and Francis.
Berridge, C. (2009) 'Modelling Pathways for a Hydrogen Fuel Infrastructure System', paper presented at the *University Transport Studies Group Annual Conference*, UCL London, January.

Bishop, S. and T. Grayling (2003) 'The Sky's the Limit – Policies for Sustainable Aviation', London, IPPR.

DTI (2006) 'The Energy Challenge', London Department for Trade and Industry.

DfT (2002) 'Powering Future Vehicles, the Government Strategy', London, Department for Transport.

DfT (2004) 'The Future of Transport: a network for 2030', Cm 6234, London, Department for Transport.

DfT (2007) 'Transport Statistics: Great Britain', London, Department for Transport.

DfT (2008) 'National Travel Survey', London, Department for Transport, www.dft.gov.uk/pgr/statistics/datatablespublications/nts/.

E4tech (2006) 'UK Carbon Reduction Potential from Technologies in the Transport Sector', Report for the UK Department for Transport, E4tech, May.

Ehrlich, P. and A. Ehrlich (1990) *The Population Explosion*, New York: Simon and Schuster.

Ekins, P., M. Hillman and R. Hutchison (1992) *Wealth Beyond Measure: An Atlas of New Economics*, London: Gaia.

EPA (1998) *G-8 Environment and Transport Futures Forum*. Report No. EPA 160-R-98-002, Washington, DC, Environmental Protection Agency.

EUCAR, CONCAWE and JRC (2002) 'Well-to Wheels Analysis of Future Automotive Fuels and Powertrains in the European Context', Brussels, CEC, December http//ies.jrc.cec.eu.int/wtw.html.

ECMT (1997) 'Internalising the Social Costs of Transport', European Council of Ministers of Transport, OECD, Paris.

Glaister, S. and D. Graham (2003) 'Transport Pricing Better for Travellers', Independent Transport Commission, Southampton.

Green Fiscal Commission (2009) 'Reducing Carbon Emissions through Transport Taxation', Briefing Paper 6, London, Green Fiscal Commission www.greenfiscalcommission.org.uk/images/uploads/gfcBriefing6_PDF_ISBN_v7.pdf.

Herring, H. and R. Roy (2007) 'Technological Innovation, Energy Efficient Design and the Rebound Effect', *Technovation* 27, 194–203.

Kwon, T. H. and J. Preston (2005) 'The Driving Force behind the Growth of Per Capita Car Driving Distances in Great Britain (1970–2000)', *Transport Reviews*, 25(4): 467–90.

Lane, B. (2004) 'The Green Car Buyer's Guide'. E book, Bristol, Ecolane.co.uk.

Maddison, D., D. Pearce, O. Johansson, E. Calthorp, T. Litman and E. Verhoef (1996) *The True Costs of Road Travel, Blueprint 5*, London: Earthscan.

Noble, B. and S. Potter (1998) 'Travel Patterns and Journey Purpose', *Transport Trends*, 3–14, London Department for Transport, Environment and the Regions.

Potter, S. (2004) 'Transport Energy and Emissions Urban Public Transport', in David Hensher and Kenneth Button (eds) *Transport and the Environment, Handbook in Transport Volume 4*, Pergamon/Elsevier, 247–62.

Potter, S. and Warren, J. (2006) 'Travelling Light, Theme 2', *T172 Working with our Environment Technology for a Sustainable Future*, Milton Keynes: The Open University.

Potter, S. (2007) 'Sustainability, Energy Conservation and Personal Transport', in S. Warren (ed.) *Managing Transport Energy Demand*, Oxford: Oxford University Press, Chapter 1.

RCEP (1994) *Transport and the Environment*, London: Royal Commission on Environmental Pollution.

Stern, N. (2006) *The Stern Review the Economics of Climate Change*, London: HM Treasury and Cabinet Office.

Wemyss, N. (1996) *Solving the Urban Transport Dilemma – the Motor Industry's Approach*, London: FT Newsletters and Management Reports.

6
A Low-Carbon Transition

Neil Strachan and Timothy J. Foxon

This chapter examines a range of techniques for analysing the technical, institutional and behavioural components of a transition to a low-carbon society (LCS). There is no one technique that answers all of the questions relative to an LCS, but we argue that the use of these techniques can both inform and complement more narrative approaches. Modellers, scenario builders and transitions analysts have argued over the last 30 years that analytical and narrative approaches can provide insights that inform policy questions, such as how to achieve a transitions to an LCS, without any one approach being able to produce definitive answers (Huntington et al., 1982). This chapter begins by discussing the definition, feasibility and achievability of an LCS, followed by examinations of the strengths and weaknesses of different analytical techniques. We examine formal energy-economic models, that can provide quantitative insights on LCS; scenarios, that can investigate broader surprises (or 'side-swipes') on the long-term transition to an LCS; and transitions pathways for an LCS, that explore alternative institutions and actors that could lead to quite different low-carbon pathways. Finally, three possible LCS transitions pathway ('Market Rules', 'Central Coordination' and 'Thousand Flowers') are outlined and key insights summarised.

Feasibility of a low-carbon society

A number of major modelling collaborative studies (e.g., van Vuuren et al., 2006; Edenhofer et al., 2006; Strachan et al., 2008; Hulme et al., 2009) have concluded that the achievement of low-carbon societies, in the context of global CO_2 emission reduction of 50 per cent by 2050 (G8, 2007), is indeed feasible in technological and economic terms.

Energy efficiency, demand-side responses, zero-carbon electricity generation and novel transport technologies are commonly identified as the most important and economically-viable contributors to emissions reduction. These studies find that the aggregate economic implications – of the order of one to two per cent of annual GDP by 2050 – are modest, although the costs would fall on particular industries and consumer groups (i.e. the old, the poor, and the rural), while new low-carbon industries and other groups (such as younger, richer, urban consumers) may benefit overall. The required carbon price signal (or marginal cost of abatement) is generally in the range of $100–350/$tCO_2$, but could be much higher in the cases of key options being unavailable. This greatly exceeds the current price of carbon in the EU Emissions Trading Scheme, which was trading at around $30/$tCO_2$ in 2008, falling to around $20/$tCO_2$ in 2009 following the impact of the economic recession (CCC, 2010: 41), indicating that current policy frameworks are not providing sufficient incentives for a low-carbon transition.

There are serious questions about the political viability of establishing a carbon price signal (through either a trading scheme or taxation) at these high levels right across the economy. Questions have been raised over the social and environmental acceptability and cost reduction potential of key decarbonisation technologies, including nuclear power, large-scale wind, sustainable biomass, fuel cells, and carbon capture and storage. There is a set of issues regarding international cooperation in emissions reductions, including flexible and equitable linking of emissions markets. There are huge uncertainties in the role of behavioural change with the willingness of current and future generations to change established energy-use and mobility practices. Finally, the institutional mechanisms and vested interests of various actors is a key uncertainty in moving forward to an LCS. This suggests that a wide range of different analytical and narrative techniques will be needed to help understand how a low-carbon transition could occur in practice and to inform decision-making by all relevant actors.

Achievability of low-carbon societies

To achieve a transition to an LCS, although achieving progress in the efficiency and affordability of a range of low-carbon technologies will be vital, changes must go to a deeper social level if climate change and development goals are to be reconciled. Key conclusions from an international modelling comparison on achieving a sustainable LCS (Skea and Nishioka, 2008), include:

- Achieving the transition to an LCS is essential if GHG concentrations in the atmosphere are to be stabilised at a safe level. Modelling and scenario work has shown that this transition is possible.
- Long-term certainty is needed to create the market conditions for investment in low-carbon solutions. This includes policy signals to strengthen carbon pricing, a comprehensive approach to RD&D for low-carbon technology, and enhanced technology transfer to developing countries.
- There are major synergies between policies that promote sustainable development objectives and those that encourage the transition to an LCS. Pursuing these policies can deliver significant economic, social and environmental co-benefits, especially in developing countries.
- The role of government is critical and top-level political leadership will be essential. Governments must establish the enabling conditions under which individuals, business and organisations can benefit from the opportunities offered by new low-carbon markets, technologies, products and services. A portfolio of policies will be required to achieve this.
- Energy efficiency improvement should be accelerated, using incentives that encourage institutional and behavioural change.
- Consumer choice and individual action, in the context of clear policies that enable low-carbon options and lifestyles, can be powerful drivers in delivering the level of behaviour change required to enable the transition to LCS.

None of this will be easy. Much further analytical work is required to articulate policy measures in the necessary detail, and to help plan through the inevitable uncertainties in such a transition. Next we review the strengths and limitations of three main analytical tools that have been used to provide insights into a low-carbon transition: energy-economic modelling; scenario development; and transition pathways.

Energy-economic modelling of low-carbon societies

In the extensive literature on energy-economic modelling of energy and climate policies, there are two widespread modelling approaches: bottom-up vs. top-down models. Bottom up models have much more emphasis on technological details of the energy system, while top-down models focus on economic links between the energy system and the wider economy (Bohringer and Rutherford, 2007). Such a typology is at best an abstraction, and many models have elements of both bottom-up and top-down as well as more detail on the micro-economic and behavioural implications of energy use. Furthermore, other model typologies can be developed; for example on whether the model is at a global, national or regional geographic level, what time frame it focuses on, how it treats uncertainty. The specific strengths and weaknesses of various approaches explain the wide range of hybrid modelling efforts that combine technological explicitness of bottom-up models with the economic comprehensiveness of top-down models (Hourcade et al., 2006).

However, in line with a 30-year old plea from the energy modelling community, models deliver insights instead of answers (Huntington et al., 1982). To provide policy makers with a sufficient scope of insights implies a range of energy models that operate at different scales and answer different questions. This key point suggests the need for strength-in-depth in different types of energy models, and the consideration of soft-linking energy models (as in the UK Energy Research Centre's Energy 2050 analysis (UKERC, 2009)) to benefit from complementary analytical strengths. This means that the outputs of one model are used as inputs to another type of model.

Over the last decade, energy system modelling has played a key role in helping UK policy makers to assess the costs, trade-offs and pathways related to achieving long-term emission targets and energy security (Strachan et al., 2009). As a key example, the iterative development of the UK MARKAL family of models have provided analytical insights underpinning the 2003 and 2007 Energy White Papers, the 2008 Climate Change Bill, the 2009 Low Carbon Transition Plan (DECC, 2009) and the ongoing reports of the Climate Change Committee (CCC, 2010). Principal insights from this body of work include that achieving a low-carbon energy system is technically and economically feasible, with multiple potential pathways to a low-carbon economy. Zero-carbon electricity and its linkages to the buildings and transport sector is a key low energy transition. But a balanced approach is required between deployment of low-carbon technologies (with increased commitment to RD&D, the strengthening of financial incentives and the dismantling of regulatory and market barriers) and behavioural change (in the uptake of existing and cost-effective energy efficiency and conservation measures as well as demand reductions in the provision of energy services).

Scenarios for low-carbon societies

A second analytical tool for assessing transitions is the development and quantification of scenarios for the assessment of key uncertainties in low-carbon societies. Recent years have seen a proliferation of low-carbon policy related studies employing the concept of scenarios. The application of scenarios in low-carbon studies has varied from providing ranges of future GHG emission levels that drive integrated assessment and other models (Nakicenovic and Swart, 2000), to long-term national energy policy analysis and initiatives (e.g. Berkhout et al., 1999). Comprehensive reviews of UK energy scenarios are detailed in Hughes and Strachan (2010) and ERP (2010).

Scenario thinking can be described as the use of the imagination to consider possible alternative future situations, as they may evolve from the present, with a view to informing and improving decisions that must be made while the future remains uncertain or undecided. The wide variety of users, and the contexts in which scenarios have been used, has led to a wide variety of approaches and methodologies towards constructing scenarios. However a broad assessment of the objectives of a range of historical strategic scenario planning activities can be encompassed by the following key objectives (Hughes, 2009):

- Protective decision making – to improve robustness to future external events;
- Proactive decision making – to improve the ability to see opportunities to actively shape future events in a beneficial way;
- Consensus building – to improve understanding between multiple actors in identifying a mutually desirable path forward.

In any given scenario study, the balance between these objectives is strongly affected by the identity of the scenario 'user' – that is the actor or actors who are intending to use the scenarios to improve their decision making. For LCS scenarios, an individual company may be interested in protective decision making, whereas a major economy government would be interested in proactive decision making and/or in building consensus among various key stakeholders.

In terms of best practice, the following guiding principles (Hughes 2009) for scenario analysis are to:

- Identify the scenario user and the range of other actors. Identify the motivations and agency of the various system actors, and the networks between them;

- Ground the scenarios in the present and demonstrate a plausible link from the present to the future;
- Challenge existing 'world-views' or 'mind-maps';
- Integrate use of both quantitative and qualitative tools and information;
- Communicate scenarios to potential users by providing a new language with which to understand the future.

Scenarios for LCS face particular challenges. First, the time frames being considered frequently extend to 50 years and beyond. Second, the subject of decarbonisation and LCS requires detailed questions on technological change. Third, LCS are also likely to be a driver for very significant social changes the like of which it may be difficult to comprehend in advance. Moreover, it is clear that dynamic social and technological changes are highly intertwined, and feedback upon each other (Geels, 2005). Fourth, LCS entail a huge range of activities within a highly complex 'socio-technical system'.

Despite these complexities, the key objectives of LCS scenarios can still be covered by the three objectives defined above; protective decisions to mitigate risks (e.g. of technology failure); proactive decisions to seize opportunities (e.g. policies to accelerate behaviour change or technological development); consensus building to engage all actors required to play a role in order to achieve success (e.g. energy companies, regulators, infrastructure operators, industrial actors, citizens, etc).

Reviews (e.g. Hughes and Strachan, 2010) illustrate strengths and weaknesses of different types of LCS scenarios applied to the UK. Trend based studies usually create a set of four scenarios, based on two alternative outcomes for two different social trends, such as globalisation vs. localisation, or 'consumerist' vs. 'community' values. These allow for more detailed speculation in areas of future social and cultural experience, often aiming to provide very vivid images of how people would live and interact with each other and with technologies in the various futures. Technical feasibility studies allow a highly controlled theoretical manipulation of physical energy systems, according to a range of imposed criteria. They are used as flexible 'testing grounds' for the theoretical plausibility of energy systems. Modelling studies combine technological detail with data on resource availability and trade-offs, global trade and interactions, and wider economic issues.

The majority of LCS scenarios focus on technical aspects, which is unsurprising given the technological complexity of energy systems, but does mean that there is far less integration of social or economic considerations. Three common major weaknesses of LCS scenario studies are

- The exogenous emissions constraint, which is assumed to be met. This underestimates a range of potentially highly significant obstacles including the interaction of actors.
- Reliance on high level trends, which although give scenario consistency, are often polarising and extreme cases (e.g., 'consumerist' vs. 'community' values), and mean that the development of these state of the world from the present is not made clear.
- How the co-evolution of social, technical and political changes is treated, including 'policy' that is often treated as an external assumption, rather than something integral to the scenario.

Despite these weaknesses, scenarios do offer a valuable tool for anticipating and exploring the unexpected uncertainties in long-term LCS pathways.

Transition pathways for low-carbon societies

A third analytical approach to LCS futures is that of transitions pathways. This builds on research in socio-technical systems, using an analytical framework based on interactions between three 'levels': technological niches, socio-technical regimes, and landscapes (Elzen et al., 2004; Geels, 2002, 2005):

- The landscape represents the broader political, social and cultural values and institutions that form the deep structural relationships of a society and only change slowly. Key landscape changes affecting energy systems are awareness and willingness to respond to climate change, fears about the security of imported energy supplies, and concerns about the affordability of energy services, particularly for people living in 'fuel poverty'.
- The socio-technical regime reflects the prevailing set of routines or practices used by actors, which create and reinforce a particular technological system. Here, the socio-technical regime for electricity reflects the current dominant technologies and systems for supplying and consuming energy, including the centralised electricity supply system based largely on coal, gas and nuclear power generation, the associated tariff system based on units (kWh) of electricity used, and the dominant role of six large, vertically integrated energy supply companies.

Because of the close connections between these technical and social elements of the system, the existing regime has high inertia and so generates only innovation leading to incremental changes in the system.

- Niches provide places for learning processes and radical innovation to occur, and are at least partially insulated from 'normal' market selection in the regime, for example, specialised sectors or market locations. Because of the fact that current supply markets are based on selling units of electricity or gas which are indistinguishable from each other, it is difficult for this type of niche to arise without structural changes to energy markets. However, some new entrant companies are aiming to sell energy services in local markets, i.e. providing a package of delivering lighting, heating or power services together with technologies or improvements, such as insulation, to improve the efficiency of these services. In relation to electricity generation, government support schemes, such as the Renewables Obligation, aim to create a protected niche market for renewable generation, so that the costs can reduce as learning occurs.

Transition pathways explore how system changes could occur through the dynamic interaction of technological and social factors at these different levels. This aims to enable consideration of at least some of these interactions, within a clear analytical framework, informed by how similar system changes have occurred in the past.

In work being undertaken by the authors and colleagues (Foxon et al., 2010), we are developing and analysing potential transition pathways to a low-carbon electricity system in the UK. This draws on the above analysis of the strengths and limitations of existing energy scenario approaches (Hughes, 2009; Hughes and Strachan, 2010), as well as by previous work on socio-technical scenarios using the multi-level framework (Elzen et al., 2004; Hofman et al., 2004). The work seeks to understand and help facilitate transition pathways for the UK, by combining historical and scenario analysis with assessment of the technical feasibility and social acceptability of potential pathways, within a whole systems assessment framework (Foxon et al., 2010). The transition pathways examine how the UK electricity system could develop under alternative plausible governance patterns and are used to analyse how these patterns could affect technological, institutional and social changes.

Figure 6.1 gives a graphical interpretation of the forces at play in a UK electricity transition pathway in moving from one socio-technical regime (the current system) to an alternate socio-technical regime (within an LCS). The future low-carbon energy regime likely will be composed of similar elements (which is why they are present in both parts of Figure 6.1), but critically can be in radically different forms and combinations, for example, with renewable and other low-carbon generation technologies replacing current fossil fuel generation technologies. (This is represented symbolically in Figure 6.1 by the different shading of the elements in the new regime.) As described above, the energy regime is composed of closely connected technological, infrastructure, behavioural and regulatory elements, which only change slowly in the absence of external drivers. However, as illustrated in Figure 6.1, the existing energy regime is destabilised by a range of landscape pressures (international, environmental, cultural), and by the development of a range of niches (social and technical) that allow experimentation of new ways of realising and meeting energy service demands. Hence this gives an evolution to a new low-carbon energy regime.

LCS transition pathways have two key advantages over conventional scenario analysis. The first is the richer representation of socio-technical systems as 'co-evolving' developments in policy, social change and technology, enables the potential for accounting for these changes in the scenario or pathway development process, rather than relying on external trends or technological extrapolations. In particular, policy is not assumed to be an external 'one-way' process, which can be switched on or 'ramped up' to deliver any level of change that a policy maker may

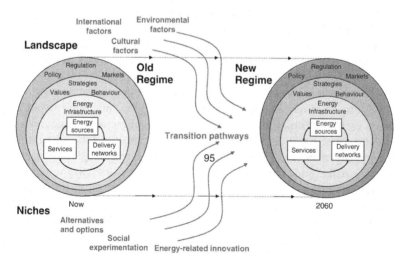

Figure 6.1 Possible transition pathways and the factors that influence them
Source: Foxon et al., 2010.

desire. In reality, existing social and technological contexts constrain and set boundaries for what may be achieved through policy – what is perceived as acceptable to society, as well as technologically possible, are real and binding constraints. Ongoing policy processes proceed iteratively, with successive implementations gradually extending this frame and delivering space for greater gains subsequently. As well as each having an iterative relationship with policy implementation, changes in technology and behaviour cannot be considered as isolated from each other. Societal forces have a huge effect in forming and conditioning the development of technologies in numerous ways (for example, public objections to new energy installations with higher costs or significant additional (e.g., visual) impacts; or the needs and interests of particular societal groups creating technological niches to incubate and eventually drive through new technologies). Equally, dynamics can flow the other way, with established technologies influencing and in some ways constraining social practices. All of the above suggests that far from being discrete entities connected only by limited linear flows; policy, technology and social change are instead bound up in an iterative or 'co-evolving' relationship (Figure 6.2).

Analysis of renewables deployment in three European countries highlighted a process of co-evolution between technological systems, institutional frameworks and business strategies that influenced the rate and form of the take-up of wind energy (Stenzel and Frenzel, 2007). Relatively small initial differences in institutional contexts in the three countries led to incumbent utilities pursuing very different strategies and hence to different patterns of deployment by incumbents and new entrants. For example, in Spain, a supportive institutional framework in the form of a 'feed-in' tariff system providing price support for renewables give rise to selective pressure for investment in wind farms by incumbents, development of relevant technological capabilities and supportive lobbying by these firms and so to high levels of wind energy deployment. On the other hand, in Germany, incumbent utilities lobbied against the introduction of a 'feed-in' tariff, and wind energy deployment was led by smaller and more local new entrant companies, who were able to take advantage of the niche created by the feed-in tariff. In the UK, the more market-oriented NFFO and Renewables Obligation support mechanisms and local planning objections resulted in a more cautious approach by incumbents and lack of incentives for new entrants, leading to much lower levels of wind energy deployment.

The second overall advantage of a transitions approach is a clear focus on actors – their motivations, and the networks between them.

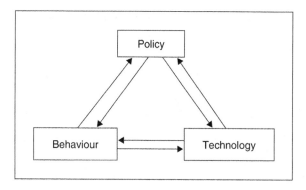

Figure 6.2 A co-evolving model of the relationship between policy, behaviour change and technology
Source: Hughes and Strachan, 2010.

Socio-technical changes are not brought about by disembodied high-level trends, or by self-assembling technical infrastructures, but by the combination of decisions and actions taken by the numerous actors in the system. Some of such decisions may be reinforcing, while others may be conflicting.

However, despite the strengths of a systemic, co-evolutionary view of transitions, such an effective descriptive present severe challenges for practical policy as there are no simple implementation mechanisms (for example compared to a CO_2 emissions price driver from modelling studies). Secondly, working from a system level perspective makes it difficult for particular actors to interpret their role in the process (Geels and Schot, 2010).

But a good transition pathway process builds a picture of a future LCS via focusing on iterative actor interactions, the dynamics of socio-technical change, considering branching points in any transition and in maintaining a systematic categorisation of different levels and types of uncertainty. This includes:

• Describing how the system may evolve – by considering an evolution of current actors networks and socio-technical interactions, a possible future can be identified via consideration of what needs to happen, by when these actions need to happen, and who could combine to undertake these actions (actors and external participants).
• Looking for branching points – in order for LCS transitions to be strategically useful, branching points (or 'points of fulcrum') where

future events can be changes corresponding to actors' key decisions, which need to be considered on a consistent basis for knock-on effects throughout the socio-technical system.

• Assessing the key uncertainties – although a grounding in present realities and a transition based on a plausible evolution of events should rule out more possibilities than one which simply postulates a range of futures.

Insights from low-carbon transition pathways

The transition pathways work is also informed by long-term historical analysis of past drivers and barriers to change in UK energy systems (Fouquet and Pearson, 2006; Fouquet, 2008). The long-term perspective shows that dramatic changes that have occurred in the mix of primary energy sources, technologies and organisational forms have led to steep reductions in the cost of energy services, and large increases in the level and range of energy service demands being met. More recent UK history shows that a key aspect of co-evolutionary change is changes to the governance of energy systems. This has ranged from a large number of small, local energy firms in the early part of the twentieth century to the dominance of large nationalised coal, electricity and gas providers from 1945 to 1979 to the current market-led approach of private firms and competitive markets for generation and supply. A key act in the early history of the UK electricity system was the creation in 1926 of the Central Electricity Board (CEB) as a panel of engineers and businessmen, independent of government, with the power to direct the evolution of the sector from a whole-systems level perspective (Hughes, 2008). This led to the creation of the National Grid, which enabled both the greater use of centralised hydro-electric and coal power supply, and the opportunity for rapid expansion of electricity-using devices, such as electric motors and domestic washing machines, in industry and homes.

Questions are now being raised as to whether the current market-led approach is compatible with delivering the UK's carbon emission reduction targets and maintaining security of energy supplies (Mitchell, 2007; Scrase and McKerron, 2009). Interesting, even the UK energy regulator Ofgem is beginning to question whether changes to current electricity and gas market arrangements will be needed to meet energy security concerns (Ofgem, 2010). The UK electricity and gas markets are currently dominated by six large, vertically integrated energy companies that supply over 90 per cent of households and industry. These companies typically pursue a strategy of investment in a portfolio of large centralised generation options, including gas-fired generation and large offshore wind power, and are active in lobbying for further institutional changes and government support for investment in new nuclear power and technologies for capture and storage of carbon emissions (CCS) from coal power stations.

With this is mind, the transition pathways work is exploring pathways under different governance patterns relating to the mix and balance of actions led by central government, actors in liberalised markets and civil society actors. The first pathway, *Market Rules*, envisions the broad continuation of the current market-led governance pattern. The government specifies the high level goals of the system and sets up the broad institutional structures, but these are based around minimal possible interference in market arrangements, which are held to be the most effective and efficient mechanism for delivering energy services. This leads to a continuation of the present centralised generation system with energy services mainly supplied by large, vertically integrated firms. The second, pathway, the government-led *Central Co-ordination*, envisions greater direct governmental involvement in the governance of energy systems. This could involve the setting up of a Strategic Energy Authority and/or the use of central contracts for delivering new low-carbon generation, including nuclear power, offshore wind and coal with CCS. The third pathway, the civil society-led *Thousand Flowers*, envisions a greater focus on more local, bottom-up diversity of solutions ('let a thousand flowers bloom'). This is driven by innovative local authorities and citizens groups, such as the Transition Towns movement (Hopkins, 2008), to develop local micro-grids and energy service companies.

Initial analysis shows that UK electricity system could evolve in quite different ways under these different pathways, but that the civil society-led pathway would lead to a more radically different energy future, with greater acceptance of reductions in the level of energy use under changed shared assumptions of what constitutes a higher quality of life. This would also lead to higher penetrations of renewable energy technologies and a greater level of decentralised energy service provision. In contrast, both the market-led and government-led pathways would lead to a greater focus on the large scale generation technologies of nuclear power, coal and gas with carbon capture and storage (CCS) and offshore wind, and incentives for technological improvements in energy efficiency, such as efficiency standards for end-use buildings' devices (e.g. appliances, space and water heating in the home). All pathways would need more active management of electricity networks through the application of information and communication technologies to develop a 'smart grid'.

So, in the government-led and market-led pathways, large energy companies see a 'high-electric' future as a strategic business opportunity, with increasing demand for electric heating and electric vehicles in a carbon-constrained world. Consumers remain in essentially 'passive' role, with initial opposition to high levels of new build of large-scale generation increasingly tempered by security of supply ('keeping the lights on') concerns and gradual widespread acceptance of climate change. In these pathways, consumers are unwilling to accept significant lifestyle changes, beyond buying more efficient appliances. Hence, the ability of the pathways to reach the 80 per cent carbon reduction target by 2050 depends largely on the key technologies of carbon capture and storage (CCS) from coal and gas-fired generation, new nuclear power stations and large-scale renewables proving to be technologically and economically feasible, as well as gaining wider public acceptance. In contrast, in the civil society-led pathway, there is a greater diversity in the range of generation technologies, including high levels of local renewable generation, in the number and scale of energy service companies, and in the willingness of householders to change their practices of energy use, such as reducing the number of electronic gadgets they have. However, this would require large-scale social changes to increase the 'visibility' and awareness of energy use by households, and also government regulation, such as to significantly increase the energy efficiency of the existing building stock, as well as new houses, through insulation and other measures. This could include (Nye et al., 2010):

- Facilitating deliberate energy conservation measures through changes in the visibility of energy, for example, the use of smart meter visual energy displays;
- Changes in habits/routines or shift to more sustainable lifestyles, such as reduction of car use;
- Changes in shared understandings of 'proper' energy use, for example, in socially-acceptable frequency of taking showers;
- Increasing demand for, and new uses for, low-carbon/more-efficient technologies (potentially leading to rebound effects);
- Increasing political action to ensure that any limitations on energy use are shared equitably between different groups.

The pathways also illustrate both the challenge of, and need for, taking a long-term perspective, as infrastructure and investment decisions made now will affect the nature of energy systems for at least the next 40 years. They illustrate that pathways arise as a result of decisions made by actors within the system, which are both constrained and enabled by existing technological and institutional structures, so that a low-carbon future will require changes to those structures, as well as individual choices.

Conclusions

This chapter has examined the strengths and limitations of three approaches that have been used to explore a transition to a low-carbon society – energy-economic modelling; scenario development; and transition pathways – and reviewed some of the insights that have come out of work following these approaches. This highlights the value of pursuing a range of approaches, as different approaches yield different insights. Energy-economic modelling can provide analytical rigour, provided the focus of the model type is appropriate to the questions being asked and the data assumptions are available and transparent, but is generally not able to deal so well with non-marginal changes. Scenario analysis brings a greater freedom to investigate alternate worlds and shocks and sideswipes that might affect the future. Scenarios can play a useful role in imagining possible LCS futures, and testing out some of the key questions of technical and social feasibility, but they are often limited by assumptions of broad trends and exogenous constraints. Transition pathways are a more recent approach, which focuses on how pathways are influenced by choices by actors, and examines co-evolutionary processes of technological developments, institutional changes, business strategies and social practices.

Studies using all of these approaches argue that low-carbon societies are feasible, but will require massive technological, institutional and behavioural changes to be brought about. They highlight the value of long-term thinking in informing current decision-making. They show that the path is not pre-determined, and that a range of high carbon or low-carbon futures are possible, depending on choices made by governments, businesses, community groups and individuals in the next few years.

Note

This work draws on research under the 'Transition Pathways to a Low Carbon Economy' project [Grant EP/F022832/1], jointly supported by the UK Engineering and Physical Sciences Research Council (EPSRC) and E.ON UK. The authors are grateful to these sponsors, as well as for the interchanges with their main UK academic partners at the Universities of Bath, Cardiff, East Anglia, Imperial College London, Leeds, Loughborough, Strathclyde, Surrey, and University College London. See www.lowcarbonpathways.org.uk.

References

Berkhout, F., J. Skea, and M. Eames (1999) 'Environmental Futures', London, Office of Science and Technology/Foresight Programme.

Bohringer, C. and T. Rutherford (2007) 'Combining Bottom-Up and Top-Down', *Energy Economics*, 30(2): 574–96.

CCC (2010) 'Meeting Carbon Budgets – Ensuring a Low-Carbon Recovery', Second progress report of the Committee on Climate Change, London, www.theccc.org.uk.

DECC (2009) 'The UK Low Carbon Transition Plan', London, Department of Energy and Climate Change, www.decc.gov.uk.

Edenhofer, O., K. Lessmann, C. Kemfert, M. Grubb, and J. Koehler (2006) 'Endogenous Technological Change and the Economics of Atmospheric Stabilization: Synthesis Report from the Innovation Modelling Comparison Project', *Energy Journal*, Special Issue 1.

Elzen, B., F. Geels, P. Hofman, and K. Green (2004) 'Socio-Technical Scenarios as a Tool for Transition Policy: An Example from the Traffic and Transport Domain', in B. Elzen, F. W. Geels, and K. Green (eds) *System Innovation and the Transition to Sustainability – Theory, Evidence and Policy*, Edward Elgar, pp. 251–81.

ERP (2010) 'Energy innovation milestones to 2050,' Energy Research Partnership report, (March 2010), www.energyresearchpartnership.org.uk.

Fouquet, R. and P. Pearson (2006) 'Seven Centuries of Energy Services: The Price and Use of Light in the United Kingdom (1300–2000)', *The Energy Journal*, 27(1): 139–77.

Fouquet, R. (2008) *Heat, Power and Light: Revolutions in Energy Services*, Cheltenham: Edward Elgar.

Foxon, T., G. Hammond, and P. Pearson (2010) 'Developing Transition Pathways for a Low Carbon Electricity System in the UK', *Technological Forecasting and Social Change*, 77: 1203–13.

G8 Communiqué (2007) 'Chair's Summary', G8 Heiligendamm Summit, 8 June 2007, www.g-8.de/Webs/G8/EN/G8Summit/SummitDocuments/summit-documents.htm.

Geels, F. (2005) *Technological Transitions and System Innovations: A Co-Evolutionary and Socio-Technical Analysis*, Cheltenham: Edward Elgar.

Geels, F. (2002) 'Technological Transitions as Evolutionary Reconfiguration Processes: A Multi-Level Perspective and a Case-Study', Research Policy, 31: 1257–74.

Geels, F. and Schot, J. (2010) 'The Dynamics of Socio-Technical Transitions: A Socio-Technical Perspective', in J. Grin, J. Rotmans and J. Schot (eds), *Transitions to Sustainable Development – New Directions in the Study of Long Term Transformative Change*, London: Routledge.

Hofman, P., B. Elzen and F. Geels (2004) 'Socio-Technical Scenarios as a New Tool to Explore System Innovations: Co-Evolution of Technology and Society in the Netherlands' Energy System', *Innovation: Management, Policy and Practice*, 6(2): 344–60.

Hopkins, R. (2008) *The Transition Handbook: From Oil Dependency to Local Resilience*, Totnes: Green Books, http://www.transitiontowns.org.

Hourcade J., M. Jaccard, C. Bataille, and F. Ghersi (2006) 'Hybrid Modelling: New Answers to Old Challenges', *The Energy Journal*, Special Issue 2: 1–11.

Hughes N. (2008) 'The Role of Public Policy and its Relation to Private Sector Investment in Driving Innovation for a Low Carbon and Secure UK Electricity System', paper for 2008 Conference of the British Institute for Energy Economics.

Hughes N. (2009) 'A Historical Overview of Strategic Scenario Planning, and Lessons for Undertaking Low Carbon Energy Policy', a joint working paper of the EON/EPSRC Transition Pathways Project and the UKERC, www.ukerc.ac.uk.

Hughes, N. and N. Strachan (2010) 'Methodological Review of UK and International Low Carbon Scenarios', *Energy Policy*, 38(10): 6056–65.

Huntington H., J. Weyant, and J. Sweeney (1982) 'Modelling for Insights, not Numbers: The Experiences of the Energy Modelling Forum', *Omega*; 10(5): 449–62.

Hulme, M., H. Neufeldt, H. Colyer, and A. Ritchie (eds) (2009) 'Adaptation and Mitigation Strategies: Supporting European Climate Policy', final report from the ADAM project, Tyndall Centre for Climate Change Research.

Mitchell, C. (2007) *The Political Economy of Sustainable Energy*, Houndsmills, Basingstoke: Palgrave Macmillan.

Nakicenovic, N. and R. Swart (eds) (2000) *Special Report on Emissions Scenarios*, Cambridge: Cambridge University Press.

Nye, N., Whitmarsh, L. and Foxon, T. J. (2010) 'Socio-Psychological Perspectives on the Active Roles of Domestic Actors in Transition to a Lower Carbon Electricity Economy', *Environment and Planning A*, 42: 697–714.

Ofgem (2010) 'Project Discovery: Final Report', London, Office of Gas and Electricity Markets, www.ofgem.gov.uk/markets/whlmkts/discovery/Pages/ProjectDiscovery.aspx.

Scrase, J. and G. McKerron (eds) (2009) *Energy Policy for the Future: A New Agenda*, Houndsmills, Basingstoke: Palgrave Macmillan.

Skea, J. and S. Nishioka (2008) 'Policies and practices for a low-carbon society', *Climate Policy*, 8: S5–S16.

Stenzel, T. and A. Frenzel (2007) 'Regulating Technological Change—the Strategic Reactions of Utility Companies towards Subsidy Policies in the German, Spanish and UK Electricity Markets', *Energy Policy*, 36 (7): 2645–57.

Strachan N., T. Foxon, and J. Fujino (2008) 'Policy Implications from Modelling Long-Term Scenarios for Low Carbon Societies', *Climate Policy*, S17–S29.

Strachan, N., S. Pye, and N. Hughes (2009) 'The Iterative Contribution and Relevance of Modelling to UK Energy Policy', *Energy Policy*, 37: 850–60.

UKERC (2009) 'Energy 2050 Synthesis Report', London, UK Energy Research Centre, www.ukerc.ac.uk.

Van Vuuren, D., J. Weyant, and F. de la Chesnaye (2006) 'Multi-Gas Scenarios to Stabilize Radiative Forcing', *Energy Economics*, 28(1): 102–20.

Part II
Case Studies

7

Case Studies in Low-Carbon Living

Robin Roy

Over two-thirds of greenhouse gas emissions in industrialised countries arise directly or indirectly from households. In Britain, homes use nearly a third of delivered energy and produce about a quarter of total CO_2 emissions (CLG, 2006). If personal transport is included the figure rises to about 40 per cent of total CO_2 emissions. If consumption of food, goods and services (including imports) is added, UK households are responsible for three-quarters of national carbon dioxide equivalent emissions. The remaining emissions are due to government expenditure (11%) and capital investment (13%) (Druckman and Jackson, 2009).

There have been a number of studies of personal and household carbon footprints. The Carbon Trust (2006) analysed the carbon dioxide emissions of an average UK inhabitant (totalling 11.3 tonnes CO_2 per year). The analysis included the personal footprint arising from UK imports and exports, and attributed the emissions from UK government activities and capital investment to individuals, but did not count the equivalent carbon footprint of non- CO_2 greenhouse gases. Using the Carbon Trust data, Roy (2009) provided the following breakdown of a UK inhabitant's carbon footprint (Figure 7.1).

Druckman and Jackson (2010) did a similar consumption-based analysis of the carbon footprint of an average UK household (totalling 26 tonnes CO_2e per year) which produced a similar breakdown, although this study included the CO_2 equivalent effects of other greenhouse gases and excluded government activities and capital investment. These UK studies and those carried out in other countries (e.g. a US analysis by Brower and Leon (1999) have found that

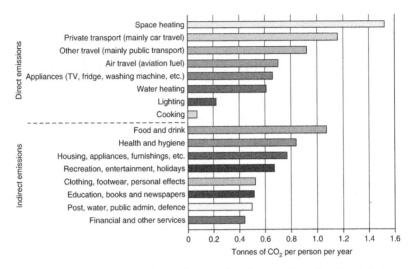

Figure 7.1   Breakdown of all the direct and indirect carbon dioxide emissions arising from the energy use and consumption of an average UK citizen
Source: Roy, 2009, p. 119, based on data in Carbon Trust, 2006, pp. 19, 24.

the main elements of an average carbon footprint in industrialised societies are:

- Transport (especially personal car travel and holiday air travel) – about 24 per cent of the total footprint;
- Domestic space and water heating – about 20 per cent of the total;
- Provision and use of services (insurance, finance, medical, recreation, hotels, education, telephone, etc.) – about 18 per cent;
- Food and drink (including agriculture, processing and distribution, catering, food imports and exports) – about 12 per cent;
- Purchases of consumer goods (clothes, shoes, domestic appliances, furnishings, books, newspapers, etc., etc.) – about 12 per cent;
- Domestic electricity use (for powering lights, appliances, etc.) – about 8 per cent;
- Housing construction and maintenance – about 6 per cent.

Eco-renovations of existing homes

This chapter provides six case studies of attempts by pioneering households, small communities and housing providers to enable a low-carbon life in Britain.

The case studies include:

- Two examples of householders improving the environmental performance of their existing home and attempting to live a low-carbon lifestyle;
- Two examples of new low-carbon homes, one designed by a small housing developer and the other by two leading green architects;
- Two examples of pioneering low to zero-carbon community housing projects.

The chapter concludes by examining the similarities and differences between the low-carbon households and communities that are illustrated by the six case studies.

Low-carbon living in an 1980s Oxford house

This case study shows what a professional couple, Matthew and Angela (not their real names), have done to eco-renovate their three-bedroom 1982 mid-terraced house in Oxford, England, as part of moving to a low-carbon lifestyle (see Figure 7.2). Below are some extracts based on their own account of what they did, plus some additional information added by the author.

Our philosophy

'Our aim is to show how a modern family can make serious reductions in their environmental impact, while still having a standard of living most of us would recognise. We aim to be practical, rather than offer solutions that few people are likely to adopt.'

Energy

We started by commissioning an energy audit of our house. This showed our house was built to low, energy efficiency standards. Our plan of action included:

1. Replacing the boiler with a modern condensing boiler and installing new heating controls and radiator valves.
2. Because of its cost, the case for installing solar water heating is usually marginal, but as we were having the central heating upgraded we decided to go for it.
3. Insulation is key in any eco-renovation. We noticed the biggest change when we had the cavity walls filled with mineral fibre. We put 100 mm sheep's wool insulation in the loft on top of the 70 mm of fibreglass, and insulated the walls and ceiling of the internal garage.

4. Our house had single-glazed windows plus secondary glazing. We decided to install the best double-glazed windows we could, and then refit the existing secondary glazing – triple glazing for a fraction of the price.
5. Building a conservatory that helps to heat the house in spring and autumn and acts as a heat buffer in winter. The conservatory was our largest single investment, costing over £20,000. It's also a great space for relaxing and growing plants.
6. Building an insulated front porch to stop cold air entering the house in winter.
7. We cook, heat and provide hot water from gas, which produces lower emissions than using electricity.

Gas and electricity consumption is about 50 per cent lower than when we first moved in. One of the lessons Matthew and Angela have learned is that planning eco-renovations is often easier than finding skilled contractors to do the work.

Water

We made an effort to save water that also reduces energy use. Nationally, slightly more than a third of all domestic water is flushed down the toilet. We chose a Swedish design, which uses two and four litres, rather than up to nine litres, a flush.

Transport

We replaced our two cars with a Toyota Prius hybrid car. It cost little more than an average saloon, about £17 500 minus a £1000 government grant available at the time, yet uses less than half the fuel of the previous cars. It has two engines, one petrol, the other electric, whose battery is charged when the petrol engine is running and by recovering energy when going downhill or braking. We get around 55 to 63 miles per gallon (5 to 4.5 litres per 100 km). Although the CO_2 emissions of the Prius are no lower than those of some super-mini diesel cars, it is a spacious family car – 'it didn't mean driving a "noddy" car, or trading in comfort or safety'. We also use our old bikes, which 'allow us to get around without always jumping into the car'. We decided to live close to the city centre and to public transport. There's no point having an eco-house and having to travel long distances to and from work. Until recently both of us could walk to work. But now Angela works eleven km away and mainly commutes by bus, except when she needs the car to travel between two sites. Matthew is changing jobs and may have to commute using their car, so Angela is exploring whether to get an electrically assisted bicycle for the days she needs personal transport.

Matthew and Angela know that air travel is a heavy polluter. But Matthew's brother lives in Canada, as do several of Angela's relatives, so they 'try to get over to that excellent country from time to time'. Matthew acknowledges that their air travel to Canada once every one to two years more than outweighs the CO_2 savings they've made from eco-renovating the house.

Food

Matthew and Angela try to reduce their food impacts by various means. These include shopping in the local covered market for fresh, unpackaged food; using a delivery service of locally grown vegetables; growing some fruit and vegetables on their allotment; buying local, ethically produced meat and eating it sparingly; and reducing their consumption of imported and out-of-season fruit and vegetables.

Goods and services

Living in a small house uses living space, and energy, efficiently. Matthew and Angela like to keep the house uncluttered, which means not buying more goods than they really need or want and recycling unwanted items. As Matthew says, 'reduce first, reuse second, and recycle third'.

Community and society

Matthew and Angela try to reduce their carbon footprint as much as they can, but recognise that there are limits to what individuals can do.

(a)   (b)   (c)

Figure 7.2 Oxford eco-renovated house (a) House Front and insulated porch (b) Rear conservatory and roof-top solar water heating panels (c) Toyota Prius hybrid petrol-electric car (Photos: Robin Roy)

They therefore help others to follow their example by providing information to friends or colleagues and by contributing to eco-renovation open days and websites. At the community level, Matthew volunteers at a local wind farm co-operative; this should supply the electricity demand of the co-op's 2500 members. Matthew and Angela also recognise that there are environmental issues that can only be tackled at government levels, such as planning, energy and transport policies.

Personal communication and interviews March 2008, in Roy, 2009, pp. 147–50.

The Yellow House

This case study shows what George Marshall, an environmental campaigner, has done to radically reduce his family's carbon footprint by eco-renovating a 1930s Oxford terraced, former council house and 'living lightly'. George Marshall bought the house in 1997 and renovated it between 1999 and 2001/2. The Yellow House – called so because of its yellow externally insulated front wall (Figure 7.4a) –gives an idea of the substantial energy savings that people living even in UK houses that are difficult to improve could make. Marshall managed his eco-renovations on a fairly limited budget using a combination of DIY and professional help.

George and his wife followed a four-stage design process:

- First they discussed 'how we wanted the house to feel and what we needed it to do'.
- They then considered the constraints and opportunities provided by existing structures and materials, the site and orientation to the sun.
- The third stage was to audit the existing energy performance of the house and identify the main areas of heat loss.
- Finally, they designed environmental improvements to meet their needs and met the building regulations.

Energy

George's aim was to reduce the household's consumption of electricity, gas and water by two-thirds compared with the previous four years. In the first year he met the goal with water consumption, and halved energy use (Figure 7.3).

To achieve the energy savings George specified about double the levels of insulation of walls, loft, floor and hot water tank than required

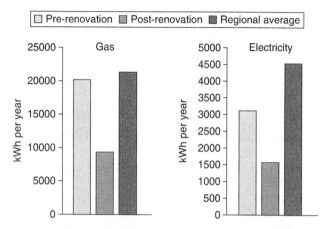

Figure 7.3 Annual gas and electricity use in the Yellow House have halved since eco-renovation. Electricity use is one-third of the regional average
Source: Roy, 2009, p. 152 based on Marshall, 2007.

by the building regulations. (Homes built or substantially altered since 2006 are reaching or exceeding these standards as a result of tighter UK building regulations.) He had most existing windows replaced with argon-filled, low-emissivity double glazing and skylights, providing the light-filled spaces they wanted.

The central heating system was replaced by a gas condensing boiler with over-sized radiators and use of controls and room temperatures to achieve optimum efficiencies. A wood stove provides occasional supplementary heat and a solar thermal system supplies most hot water, which the family attempts to use efficiently, e.g. showering after the sun has heated the water.

George replaced the old rear extension with a new one incorporating skylights plus a sun space, built partly from salvaged materials, to gather passive solar energy (Figure 7.4b).

To reduce use of electricity the refrigerator, washing machine and dishwasher are energy efficient models, while the cooker is gas. Washing machine and dishwasher are filled by partly solar heated water. George even experimented with adding external insulation to their second-hand chest freezer (ensuring that its external condenser coils were not obstructed).

Natural lighting from windows and skylights, plus glass bricks in internal walls, is used as much as possible; and any artificial lights are low-energy ones. Having stopped unwanted draughts, George designed

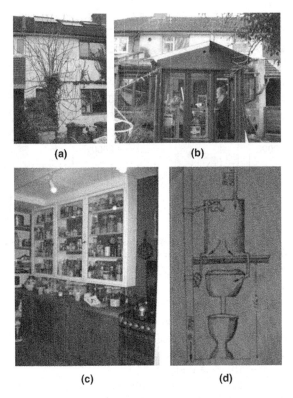

Figure 7.4   (a) Front of the Yellow House with its externally insulated concrete block walls, evacuated tube solar water heating panel and roof lights (b) The rear sunspace (c) The kitchen constructed from reclaimed furniture (d) The bathwater toilet flushing system (Photos: Robin Roy)

a system to provide natural ventilation using pre-warmed air from the sun space flowing through rooms via adjustable vents and exiting from skylight vents at the top of the house. Warm air gathering at the top of the house is sometimes drawn back down using a fan.

Materials

George is very keen on using salvaged and recycled materials to reduce waste and embodied energy and the need for environmentally damaging materials. For example, much construction timber was salvaged and the kitchen was constructed from second-hand solid wood furniture installed under a new timber worktop (Figure 7.4c). The result may not be to everyone's taste, but the recycled kitchen certainly saves money and reduces impacts.

Water

George also wanted to reduce the house's consumption of water by two-thirds. This was mainly achieved by his DIY grey water system that collects waste bath and shower water in a tank above the toilet cistern and so is used to flush the downstairs toilet (Figure 7.4d). He also installed a low flush device on the upstairs toilet and a large water butt to collect rainwater for the garden.

Transport

George Marshall says little on transport, except 'we refuse to have a car and travel by bicycle and public transport ... Managing without a car is a struggle in a world that assumes that all parents have cars'. However, since the birth of their second child, the Marshalls reluctantly bought a car, partly to be able to take their child, who suffers from asthma, to hospital at short notice. But the car is only used once or twice per month for local trips and for the occasional longer trip.

Food and waste

The family attempts to buy as much local food as possible from an organic box scheme and local farms and grows some fruit and vegetables on an allotment. They have also tried energy saving cooking methods such as using a modern 'hay box'. Almost all their organic waste is composted.

Community and society

George as part of his environmental work has produced a Yellow House website and CD (Marshall, 2002) which provides information on how to reduce environmental impacts in the above and other areas, including using recycled and salvaged materials and choosing low impact household appliances, flooring and furnishing. In 2009 the Marshalls moved from the Yellow House to eco-renovate a listed building in Wales.

(Based on Roy, 2009, pp. 151–4; further information from www. theyellowhouse.org.uk.)

New low- and zero-carbon homes

Below are examples of new low-carbon homes, one designed by a small housing developer and the other by two leading green architects.

Millennium Green

Millennium Green was one of the first developments of green homes aimed at the mainstream market. It comprises twenty four environmentally efficient houses and a business centre for residents built in 1999–2000 by a small developer, Gusto Homes, at Collingham, near Newark, Nottinghamshire.

The firm's managing director had seen examples of ecological housing and felt there was a gap in market for such homes among the ordinary house-buyer.

Materials and energy

The houses are conventional looking but incorporate many environmental features. Construction materials, such as sustainable timber, locally sourced bricks and clay drainage pipes, were chose to minimise environmental impacts, Insulation levels greatly exceeded the UK Building Regulations then in force: including 150mm insulation in walls and floors, 300mm of recycled newspaper in the roof space, and argon-filled, double glazed timber windows. Energy demand was further reduced by south-facing main glazing, mechanical ventilation with heat recovery and solar water heating systems. But despite their low heat demand the houses all have gas condensing boilers and radiators because local estate agents said that customers would not buy a house without central heating; and this view was supported by at least some of the early buyers. Also, some buyers preferred incandescent decorative lights rather than the compact fluorescent lamps provided.

Water

Water saving is one of the main features of the development. The homes have a rainwater harvesting system comprising an underground water tank large enough to supply over two weeks supply of water for toilet flushing, washing machines and gardening (Figure 7.5b). The tank is topped up from the mains when necessary and the system alerts the householder when this is occurring. Monitoring of two households showed that harvested rainwater supplied half of their water demand.

Transport

To help reduce the need for commuting, high speed internet connections were provided in the houses and an on-site business centre offers office space to enable residents to work from home. To encourage community and wildlife a shared open space and protected area have been provided.

(a)   **(b)**

Figure 7.5 (a) One of the larger houses at Millennium Green, showing main glazing, conservatory and solar water heating panel on the south side. Insulation is much greater than required in the 1999–2000 UK Building Regulations. For commercial reasons, despite its low heat requirements, the house has gas central heating. (b) Water for toilet flushing and washing machine is collected from the roof and stored in a large underground tank in the garden.

This four-bedroom house was priced in 1999 at £170 000; two-bedroom properties on the development started at £60 000. (Photo: Robin Roy)

Gusto Homes found that the Millennium Green homes cost about ten per cent extra due to their environmental features, but this did not deter buyers, many of whom were attracted by the green features.

Based on Roy (2005); additional information from Sustainable Development Commission (2005).

The Autonomous House

One of the inspirations for Millennium Green was a much more radical attempt at creating a home for low-carbon living. This was the Autonomous House designed by pioneering green architects Robert and Brenda Vale and built in 1993 in Southwell, Nottinghamshire. The Vales originally built the Autonomous House for themselves, but by 1999 they had moved and the house was rented to an environmentally conscious couple.

Materials and energy

Local and recycled materials were used extensively for construction of the house which was designed to last hundreds of years. The house is almost entirely heated from passive solar energy plus incidental gains from the occupants, appliances and lights. With super-insulation, small triple-glazed main house windows and a very thermally heavy concrete construction to store heat gains there is no need for a main heating system and the new occupants only used the small wood stove for supplementary heating four times by mid Winter. Photovoltaic (PV) panels

(a)   (b)   (c)

Figure 7.6 (a) Cross-section of the Autonomous House showing the superinsulation in walls (250 mm mineral fibre) and roof (500 mm recycled newspaper). (Brenda and Robert Vale, architects.) (b) The south-east facing upper conservatory for passive solar gain and the small windows of the main house to minimise heat loss. (c) A 2.2 kW grid-connected photovoltaic array supplies about half of the electricity demand. (Photos: Robin Roy)
Source: Based on Roy (2005), further information Vale and Vale (2000).

(Figure 7.6b) supply about half the household's electricity, mainly for water heating, lighting and cooking – no dishwasher or freezer was provided – and there is no solar water heating. Monitored energy consumption is about 13 per cent of a conventional house built to the 1995 Building Regulations (BRECSU, 1998).

Water

The house is also independent of mains water and sewerage. Rainwater from the roof is filtered through sand in the lower conservatory area and stored in tanks in the cellar. Water for drinking and cooking is further treated by a ceramic/carbon filter and drawn from a separate tap. The cellar also accommodates the tank of the composting toilet.

Low-carbon living

The total project cost of the house was fairly high at £145 000, including £15 000 for the PV array but excluding land. Also living in it requires some commitment, such as opening and closing doors and windows to maximise solar gain, accepting occasional low winter room temperatures, checking the water filters and tanks and being very careful with water use to avoid it running out. Small windows mean relatively poor natural lighting in the main house. Probably the biggest difference from a conventional house is the need to regularly rake and empty the composting toilet.

The Autonomous House shows that it is possible to design housing that achieves the required 80–90 per cent reductions in resource use and emissions, but such homes are rare and most developers and house buyers would not want to attempt such levels of sustainability.

New low-carbon communities

Next are two examples of pioneering low to zero-carbon community housing projects, one of five houses and the other much large scale for over 80 homes.

The Hockerton Housing project

One way of achieving sustainability for a greater number of people is to go beyond individual low-carbon households to communities that are environmentally, socially, and economically sustainable. A pioneering example of this approach is the Hockerton Housing Project (HHP) that was inspired by the Vales' Autonomous House but takes the concept to the small community level, thus sharing some of the cost and effort of building and living sustainably. HHP's five family homes were constructed by the builder of the Autonomous House with help from its future residents at Hockerton village near Southwell, Nottinghamshire and completed in 1998. Each household invested about £90 000 in the project.

Materials and energy

HHP's homes look distinctive because of the earth covering over the rear and the conservatories along the south-facing front, which trap solar heat from where it is let into the house, and. The interiors are light and modern and stay warm all year with no artificial heating because of the passive solar design, temperature regulating earth sheltering and thermally heavy concrete construction, plus triple glazing, super-insulation and heat recovery mechanical ventilation (see Figure 7.7). Apart from the concrete and insulation, construction materials such as bricks and timber were chosen for reduced environmental impact.

Originally hot water was supplied by a heat pump that extracted heat from the conservatory air. Following the failure of most of the heat pumps, hot water is provided by an electric heater in a large highly insulated water tank. Initial total energy use for each unit was monitored at 25 per cent of an average UK home (BRECSU, 2000). By 2002, after overcoming opposition from local residents, electricity for lighting, appliances and hot water was supplied from a five kilowatt (KW) wind turbine and a 7.6 kW photovoltaic array on the roof, both linked to the National Grid. This allowed each household's energy demand to be reduced to ten per cent of the UK average and the project to meet the 'autonomous' standard for housing with almost no net CO2 emissions. In 2004 a second five kW turbine was installed to power a new Sustainable Resource Centre.

Water

A reed bed system provides on-site sewage treatment. The bed is an area of the lake planted with bulrush and common reed. Sewage is trickled through the bed where the roots of the plants allow contact between the waste water and oxygen in the air. This creates conditions conducive for the bacteria that feed on the organic matter in the sewage and reduce its polluting potential. The resulting effluent is clean enough to discharge into the lake without causing harm. Rain from the conservatory roofs is stored in tanks, filtered and treated with ultra-violet light and pumped to the homes for drinking. Non-drinking water is collected from the surrounding land.

Food and transport

HHP residents produce over half of their food on-site, including vegetables, fruit, honey, eggs and some meat. Each household is permitted one conventional car, which they try to share, while cycles and an electric car provide local personal transport.

Community and society

HHP residents earn their living on-site, for example, providing sustainable living courses and tours while others have conventional jobs. Each household is expected to provide 300 hours per year to work for community activities such as maintenance and growing food. In other words, the project aims to be socially and economically as well as environmentally sustainable.

Despite some initial scepticism and opposition to its first wind turbine, over time the project has changed attitudes in the local area. In 2006 a group of Hockerton villagers, helped by HHP members, got together to raise finance to buy a community owned wind turbine. A second-hand 225 kW turbine was eventually installed on local farmland and in its first year of operation in 2010 generated enough electricity to nearly meet the village's total annual use.

Beddington Zero Energy Development (BedZED)

An example of a sustainable community at a bigger scale than the HHP is the BedZED (Beddington Zero Energy Development) in South London.

(a)   (b)

(c)   (d)

Figure 7.7 The Hockerton Housing project. (a/b) Five terraced units built in 1996–8 combine a superinsulated 'earth sheltered' solar energy design that provides all space heating with on-site water collection and sewage treatment and small-scale organic food production. (c) Each household can have one fossil-fuelled car, but cycles and a shared electric car are used where possible. (d) Subsequently two wind turbines and photovoltaic cells were installed to provide the much of the project's electricity, including the Sustainable Resource Centre. (Photos: Robin Roy and Horace Herring)

Source: Based on Roy (2005); further information from www.hockertonhousingproject.org. uk.

BedZED began in the mid-1990s when the director of BioRegional, an entrepreneurial sustainability charity, met green architect Bill Dunster, which led them to try to develop Dunster's idea of a zero-carbon urban community. The project was initiated by BioRegional and Bill Dunster Architects (now ZEDFactory), developed by the Peabody Trust, a large housing association that provided much of the funding, plus several partners, and completed in 2002.

BedZED houses about 200 people in 82 houses, flats and maisonettes, each with its own small ground level or roof garden, comprising about 40 per cent housing for sale, 40 per cent key worker rental and shared ownership and about 20 per cent social housing for rent.

The brightly day-lit, contemporary interiors, conservatories and gardens of the BedZED homes (Figure 7.8) and workspaces have proved very attractive with all quickly sold or rented. The rooms are rather small, but people seem prepared to pay about 15 per cent extra for BedZED's design, cost saving and green features – one-bed flats sold in 2002 at about £100,000 and four-bed town houses for £240,000 with prices changing in line with the housing market.

Materials and energy

As many as possible of the construction materials were locally sourced or reclaimed. The superinsulated terraces with south-facing solar conservatories and thermally heavy construction have minimal space heating requirements, hence no conventional central heating systems are required, only a towel rail/radiator in the bathroom. One of BedZED's most distinctive features are the brightly coloured roof-mounted room ventilators that turn into the wind, draw in fresh air and expel stale air via a heat exchanger (Figure 7.8). Some electricity is provided by PV cells on the south-facing walls and roofs.

Monitored BedZED households use less than 20 per cent of the heat and just over 50 per cent of the electricity of an average local resident (Hodge and Haltrecht, 2009). This is despite the poor performance and decommissioning of BedZED's combined heat and power (CHP) plant, which was intended to be fuelled by gas produced from tree surgery waste and to provide the community with electricity, piped hot water and space heating. Instead hot water and the small amount of room heating required is supplied from a central gas boiler and most electricity comes from the national grid.

Water

Waste water and sewage, processed through filters and a floating reed-bed in a greenhouse, is used for toilet flushing and garden use. The use of rainwater collected from the roofs has been successful, but the reed bed system was too energy intensive and costly to operate and was replaced by an experimental membrane bio-reactor. Household mains water demand has nevertheless been cut to less than half the local average (Hodge and Haltrecht, 2009).

Transport

A green transport plan was part of the project from the start. The average car mileage of residents has been cut to under half of the local average by locating the development near bus, rail and tram routes, by

(a)   (b)   (c)

Figure 7.8 BedZED, South London (a, b, c) Views of the south-facing terraces with their roof ventilators, wall-mounted PV cells, full height solar conservatories, and ground level front or 'sky' gardens reached from a first floor bridge. (Photos: Robin Roy)

Source: From Roy (2005); further information from Bioregional.com/what-we-do/our-work/bedzed.

restricting and charging for parking spaces, providing onsite workspaces and covered bicycle parking, and having a car-sharing club. However, it has been found that BedZED occupants fly more than average, so that their transport footprint is higher than an average UK resident (Hodge and Haltrecht, 2009).

Community and society

Although the main aim of BedZED is very low CO_2 emissions, it also aims to be economically sustainable via the work spaces and socially sustainable by providing social housing units, a nursery (since closed), allotments, village square, community centre and sports field.

Because of its many innovative features, the cost of the BEDZED project greatly exceeded the budget and the CHP and waste-water systems did not work properly. So, although widely admired, these problems somewhat damaged the reputation of sustainable housing among some housing professionals. Also, some local people feel that the novel design does suit the surrounding area.

Projects like BedZED show what can be achieved, while its higher costs and technical bugs are not surprising given the degree of innovation it involved. Since BedZED was completed its ZEDFactory architects have designed a timber frame kit house called RuralZED with heat-retaining concrete walls and ceilings that can be specified to meet the various levels of the UK's Code for Sustainable Homes up to the zero-carbon standard (Dunster et al., 2008). The first six zero-carbon 'One Earth Homes' based on RuralZED were completed in 2009 for a housing association as part of a development of 1360 new low-carbon homes at Upton, Northamptonshire.

Comparison between low-carbon living pioneers

All the case studies in this chapter are of individuals, communities and housing providers in Britain who are pioneering low-carbon homes and lifestyles mostly beyond the mainstream of households in industrial societies. What features do they have in common and how do they differ from each other?

Table 7.1 compares the case studies on the main elements of low-carbon living.

The table shows that the six low-carbon living pioneers share a great many features, but there are some important differences.

Energy

Something clearly shared by all the pioneers is substantial home energy saving. This is achieved in all cases by heavy insulation, thermally heavy construction and passive solar heating plus low energy lights and appliances. The Autonomous House and HHP homes are almost fully passive solar heated; the remaining households burn gas to provide some or most space heating. Hot water is supplied in various ways, almost all with a solar thermal or PV contribution. Two-thirds of the homes have heat recovery ventilation systems. Nevertheless, most of the energy technologies employed in these building are fairly well-established; when innovative or complex systems such as heat pumps and biomass CHP systems have been tried they have failed. In terms of size, only the Autonomous House and the larger Millennium homes can be considered generous for their number of occupants, while the Oxford terrace and BedZED units provide for compact living.

Water

Another common feature is water saving. Although the carbon emissions from supplying a household with mains domestic water are relatively small water conservation helps reduce the pressure on scarce resources. Two sites rely solely on harvested rainwater, two on mains plus rainwater, and two solely on mains water. To reduce consumption the Autonomous House has a composting toilet while the rest use low flush WCs; the Yellow House also has a grey-water recycling system. Half water their gardens exclusively with collected rainwater and half

	Eco-renovations (householders)		New low- and zero-carbon homes (householders, communities, housing providers)			
	1980s Oxford house	The Yellow house	Millenium Green	Autonomous House	Hockerton Housing Project	BedZED
Location	Urban	Suburban	Suburban	Urban	Rural	Urban
Developer	Householder (+contractors)	Householder (+builder)	Small housing developer	Green architects (+contractors)	Community group (+green architects, builder)	Housing association (+green architects, consultants, contractors)
Energy						
Home insulation	High	High	High	Super	Super	High
Windows	Triple (double + secondary)	Double (low e, argon)	Double (low e, argon)	Triple (low e, argon)	Double (low e, argon)	Double (low e, argon)
Passive solar	Yes (conservatory)	Yes (sun space + skylights)	Yes (orientation, windows, conservatory)	Yes (orientation, windows, conservatory)	Yes (orientation, windows, conservatory)	Yes (orientation, windows, conservatory)
Thermally heavy construction	Yes (cavity fill masonry + tiles)	Yes (ext. insulated concrete block)	Yes (cavity fill masonry + blocks)	Very (thick concrete slab + tiles)	Very (cavity fill concrete + tiles)	Yes (cavity fill masonry + blocks)
Space heating system	Gas condensing + radiators	Gas condensing + oversized radiators (+wood stve)	Gas condensing + radiators	None (occasional wood stove)	None	Central gas boiler (one radiator per home)

(continued)

Table 7.1 Continued

	Eco-renovations (householders)		New low- and zero-carbon homes (householders, communities, housing providers)			
	1980s Oxford house	The Yellow house	Millenium Green	Autonomous House	Hockerton Housing Project	BedZED
Hot water	Solar + gas	Solar + gas	Solar + gas	Solar PV + mains electric	Solar PV (+mains top-up)	Central gas boiler
Ventilation	Windows, (one heat recovery fan)	Natural stack + heat recovery	Mechanical heat recovery	Windows, natural stack	Mechanical heat recovery	Natural stack + heat recovery
Electricity supply	Grid	Grid	Grid	PV + grid	Wind + grid + PV	Grid + PV
Appliances + lights	All energy saving	All energy saving	Most energy saving	All energy saving	All energy saving	All energy saving
Compact living space	Yes	Somewhat	Varies for house sizes	No	Somewhat	Yes
Water						
Water supply	Mains	Mains + grey water toilet flush	Rainwater + mains top-up	Rainwater	Rainwater	Mains + rainwater
Toilets	V. low flush	Low flush	Low flush	Composting	Low flush	Low flush
Waste water treatment	Mains	Mains	Mains	Composting toilet, soakaway	Outdoor shared reed bed	Experimental bio-reactor
Garden/allotment	Rainwater + mains	Rainwater + mains	Rainwater + mains	Rainwater	Rainwater	Rainwater

Materials and waste						
Low impact construction materials	Some	Some	Most	Most	Most	Most
Reclaimed/salvaged materials	No	Extensive use	A few	Many	Some	Some
Waste recycling	Yes	Yes	Yes	Yes	Yes	Yes
Composting	Yes	Yes	Yes	Yes	Yes	Yes
Food						
Local food	Yes	Yes	Some	Yes	Yes	Yes
Grow own fruit/veg	Yes	Yes	Some	Yes	Yes	Some
Produce other food	No	Catch fish	No	No	Poultry, fish eggs honey	No
Organic food	Yes	Yes	Unknown	Unknown	Yes	Some
Transport						
Max. one car per household	Yes	Yes	Some households	Yes	Yes	Mostly
Low-carbon vehicles	Hybrid car, cycles	Cycles	Unknown	Unknown	Cycles, shared electric car	A few
Public transport/shared vehicles	Yes	Yes	Some?	Unknown	Car, cycle sharing,	Yes incl. car club
Home working	Partial	Partial	Partial + on-site business centre	Unknown	Partial + on-site work centre	Partial + on-site workshops, offices
Leisure air travel	Above UK average	Unknown	Yes, unknown impacts	Unknown	Unknown	Above UK average

(continued)

Table 7.1 Continued

	Eco-renovations (householders)		New low- and zero-carbon homes (householders, communities, housing providers)			
	1980s Oxford house	The Yellow house	Millenium Green	Autonomous House	Hockerton Housing Project	BedZED
Community and society						
Shared community facilities/ activities	No	No	Wildlife and community area	No	Food growing, water treatment, maintenance	Community nursery, café, sports field
Involved in promoting sustainable living	Yes	Extensively	Unknown	Yes	Extensively	Some
Some problems	Finding skilled trades to do eco-renovations; air travel trade-offs	Inadequate public transport for 2 child family	Acceptance of low energy lights by some households	Emptying composting toilet; natural lighting of main house	Failure of heat pumps	Cost overrun; Failure of CHP and waste water systems

with rainwater supplemented by mains. The Autonomous House, HHP and BedZED treat all their waste water on-site. Although some of these systems may reduce carbon emissions, recent research has shown that rainwater, grey-water and on-site waste water treatment systems often increase emissions compared to mains water supply, because of the energy embodied in equipment and electricity for pumps (NHBC, 2010). This shows that what is conventionally considered 'green' may not always prove to be the optimum environmental solution.

Materials and waste

Unsurprisingly all the low-carbon pioneers recycle household waste and compost their food and garden waste. Most of the homes are built from at least some local, low impact and/or recycled materials (usually hidden in the structure), while the Yellow House makes extensive use of salvaged items. Such choices seem to be partly a matter of whether householders are prepared for a 'recycled' rather than a glossy new look to their home; which relates to consumption values of the people concerned.

Food

Food is an area with plenty of scope for reducing emissions. Again it is not surprising to find these low-carbon pioneers attempting to reduce 'food miles' by shopping locally and growing at least some vegetables and fruit in a garden or allotment. However, only HHP residents attempt to produce items such as meat and eggs as they have land available plus a commitment to shared community work for food growing. However, even they cannot hope to be self-sufficient in food. Organic food is at best marginal in carbon saving, but several of the households buy at least some organic for reasons of health or taste.

Unfortunately there is little information about the goods and services consumption of the low-carbon pioneers even though such consumption is a major part of an individual's or household's carbon footprint. Only the Oxford house couple stated that they deliberately avoid purchasing too much, at least partly to avoid cluttering their small home, although it is likely that committed environmentalists living in the other homes would also aim for more sustainable consumption. However, on average monitored BedZED residents bought similar amounts of consumer goods as a typical UK household (Hodge and Haltrecht, 2009).

Transport

Transport is one of the most difficult areas in which to reduce consumption, but most of these pioneers are attempting to reduce their transport footprints. Nevertheless, most own a car, even if confined to one per household either from choice or at Hockerton and BedZED by regulation. Public transport, car sharing, cycles, hybrid or electric cars are also used but the choices are much more dependent on location (urban or rural) and personal circumstances than say home energy saving. For example, the Yellow House family reluctantly acquired a car in case their second child needed urgent transport to hospital Air travel in particular poses a dilemma for those attempting low-carbon living, especially since environmental pioneers tend to be wealthier than average and so can afford more travel. Although there is little information about these pioneers' air travel, there is some evidence that flight emissions are reducing at least some of the benefits of their low energy homes. The Oxford couple fly every one to two years to visit relatives in Canada and typical BedZED occupants have been found to fly more than average, annually flying the equivalent of a return trip to New York (Hodge and Haltrecht, 2009).

Diffusion of low-carbon homes and living

Since these case studies were completed there has been an increased interest in sustainable housing, stimulated by the 2007 UK government announcement that by 2016 all new homes should be 'zero-carbon'. This initially implied that over a year each new dwelling or development should produce net zero CO_2 site emissions, but in practice is likely to mean low-carbon homes with investment in off-site 'allowable solutions' such as community heating schemes, to offset any remaining on-site emissions. The strategy for achieving zero-carbon is the Code for Sustainable Homes, which sets CO_2 reduction targets, and awards points for other features including water saving, low-impact building materials, and waste recycling (CLG, 2010). The code has resulted in architects and house-builders designing prototype zero-carbon homes and housing developments. In addition, several initiatives to encourage and help householders and others to eco-renovate existing homes have emerged. These complement government incentives to promote the installation of home insulation as well as low-carbon technologies such as solar PV and heat pumps that are currently mainly confined to a niche market of green householders (Roy, Caird and Abelman, 2008). Hence, the innovations pioneered by these case study low-carbon homes and living experiments are being promoted and adopted much more widely.

However, living in a low or zero-carbon home is no guarantee of living a low-carbon lifestyle, especially if the householders are relatively wealthy, hence the need to also tackle the impacts caused by buying more electronic equipment, lights and appliances, increased travel and the globalised consumption of food, goods and services.

Note

The author would like to thank all the low-carbon living pioneers who agreed to be filmed and/or interviewed for Open University course materials and/or provided information about their homes and lifestyles, the OU/BBC and independent producers of the videos that resulted; and Stephen Potter and Horace Herring for valuable help and comments.

References

BRECSU (1998) 'Building a Sustainable Future', General Information Report 53, London, Department of Environment Transport and Regions, October 1998. www.action21.co.uk/assets/docs/construction/gir053.pdf.

BRECSU (2000) 'The Hockerton Housing Project – Design Lessons for Developers and Clients', New Practice Profile 119, Watford, Building Research Establishment.

Brower, M. and Leon, W. (1999) *The Consumer's Guide to Effective Environmental Choices*, New York: Three Rivers Press.

Caird, S., R. Roy and H. Herring (2008) 'Improving the Energy Performance of UK Households: Results from Surveys of Consumer Adoption and Use of Low and Zero Carbon Technologies', *Energy Efficiency*, 1(2): 149–66.

Carbon Trust (2006) 'The Carbon Emissions Generated in All that we Consume', London, The Carbon Trust. www.carbontrust.co.uk/Publications/pages/publicationdetail.aspx?id=CTC603.

CLG (2006) 'Review of Sustainability of Existing Buildings: The Energy Efficiency of Dwellings – Initial Analysis', London, Department for Communities and Local Government. www.communities.gov.uk/publications/planningandbuilding/reviewsustainability.

CLG (2010) 'Code for Sustainable Homes: Technical Guide', London, Department for Communities and Local Government. www.communities.gov.uk/planningandbuilding/sustainability/codesustainablehomes/.

Druckman, A. and T. Jackson (2009) 'Mapping our Carbon Responsibilities: More Key Results from the Surrey Environmental Lifestyle Mapping (SELMA) framework', RESOLVE Working Paper 02-09, Guildford, University of Surrey. www.surrey.ac.uk/resolve/Docs/WorkingPapers/RESOLVE_WP_02-09.pdf.

Druckman, A. and T. Jackson (2010) 'An Exploration into the Carbon Footprint of UK Households', RESOLVE Working Paper 02-10, Guildford, University of Surrey, 2010) www3.surrey.ac.uk/resolve/publications/index.php?pubCat_ID=4.

Dunster, W., C. Simmons and R. Gilbert (2008) *The ZEDbook: Solutions for a Shrinking World*, Abingdon: Taylor and Francis.

Hodge, J. and J. Haltrecht (2009) 'BedZED Seven Years On', Wallington, Surrey, BioRegional. www.bioregional.com/files/publications/BedZED_seven_years_on.pdf.

Marshall, G. (2002) 'The Yellow House Guide to Eco-Renovation, CD-Rom and Website', Oxford, Climate Outreach and Information Network. http://theyellowhouse.org.uk.

Marshall, G. (2007) 'The Yellow House: Eco-Renovation of a 1930s Terrace House', Presentation at Climate Change, Oxford, Climate Outreach and Information Network, 5 June 2007. www.coinet.org.uk.

NHBC (2010) 'Energy and Carbon Implications of Rainwater Harvesting and Greywater Recycling', Bristol, Environment Agency. www.nhbcfoundation.org/Researchpublications/CarbonImplicationsofRainwaterHarvesting/tabid/429/language/en-GB/Default.aspx.

Roy, R. (2005) Video 'Green Homes' and Audiovisual Notes 1, OU course T172 Working with our environment, Milton Keynes: The Open University.

Roy, R. (2009) 'Treading Lightly on the Earth', In *Block 1 Setting out From Home*, OU course U116 Environment, Milton Keynes: The Open University, pp. 107–68.

Roy, R., S. Caird S. and J. Abelman (2008) 'YIMBY Generators: Yes in my Back Yard! UK Householders Pioneering Microgeneration Heat', London: The Energy Saving Trust. http://oro.open.ac.uk/10828/.

Sustainable Development Commission (2005) 'Case Study: Millennium Green'. http://www.sd-commission.org.uk/communitiessummit/show_case_study.php/00044.html.

Vale, R. and B. Vale (2000) *The New Autonomous House*, London: Thames and Hudson.

8
Designing and Creating my Low-Carbon Home

Catherine Mitchell

By taking on the 1870s coastguard cottage in Cornwall in 2008, I was already committed to a certain level of renovation as the property was in a bad state; cold, leaky and damp. What it did offer was the perfect opportunity for me to create a low-carbon home and let me experience firsthand the process that many more homeowners will go through if there is to be an ambitious programme of low-carbon retrofitting in the UK. My main criteria were that the house be very energy efficient, modern and simple. It soon became clear that there were two ways in which this could be achieved – refurbish the old 1870s building or knock much of it down and then rebuild it. In the end, the latter was chosen because although the cost was not going to be very different, the energy efficiency of the house was going to significantly improved with the new timber frame walls.

There would be many more of these crossroads moments in the project. Attempting to refurbish or build a low-carbon house requires complex choices to be made at each stage of the design and build process. One set relates to the actual building – and skimmed over here – while the other set of choices relate to what is seen, and lived with, within the house – doors, windows, passive lighting, electrical goods, furniture, baths etc. In order to comply with building regulation, fairly major decisions about the build are required from the outset, regarding for example, onsite generation, heat provision; central heating (including whether to have radiators or under floor heating) or other forms of heat such as a wood burning stove; the choice of windows; the position of electric points and lights; whether to have an air flow exchanger and so on.

If the refurbishment is extensive enough to require planning permission, a key decision is whether or not to take on an architect. The extensive refurbishment of my own property required one, as half of the original house was to be knocked down, rebuilt with a 6 m² footprint but an extra two upstairs rooms, with complete re-wiring, re-plumbing and insulation. As with all new construction, building regulations are there to ensure that the final building complies with the plans set out in the early building design document. In theory, a good builder should be able to follow exactly a building regulation design document, but in practice I found it useful to have the project supervised by someone with experience of sustainable and energy-efficient building design.

Space heating

Wanting a warm home was the basis on which all decisions were made relating to the space heating of the house. Consequently, I have ended up with van loads of insulation, much of it sheep's wool, stuffed everywhere. Between the ceiling of my first floor and the attic floor, the insulation is 400 mm – vital for a cosy house but consequently I can no longer stand in my attic (although it is a perfectly viable crawling space). This was an easy decision for me; I recognise that for bigger families needing more storage space, this could be more of a nuisance. There is cavity wall insulation in the 'old house' walls, and external walls also had external insulation added.

Space heating is provided by a natural gas boiler and a wood burning stove (WBS). My project manager said I would not need any central heating but I was worried that there would be times when I would be cold and so insisted I had one installed – with all the palaver of pipes and radiators. But he was right. I would not again install central heating. I have used it only rarely and if I did not have it I do not think I would miss it. On the other hand, I really enjoy my WBS – and tend to use it for aesthetic effect as much as for space heating. In order to maintain the thermal comfort achieved through the comprehensive insulation effort, enormous care was taken to ensure that the wood burning stove exit pipe was installed so that it did not allow outside cold air into the house. Effectively, the entire house is 'sealed' so that heat transfer should be limited to that of the 'u' value of the walls and glass, with an air exchanger included to ensure fresh air. Creating this seal adds various conditions to a house, such as avoiding letter boxes and cat flaps. It also means that adding additional pipes that go from the inside to the outside of the house at a later date is not easy if the same level of seal is to be maintained.

As windows have such a significant impact on the thermal comfort and overall look of a property, I spent a lot of time considering the type of window to use. In the end I settled upon triple glazed Internorm windows as they offered optimum thermal insulation and soundproofing, and alongside this, are beautifully designed and easy to use. My planning permission required that the windows in the front of the 'old' part of the house looked like the original sash 1870 windows. My front and back door are also triple glazed, only a passive lighting ceiling area is double glazed for reasons due to sourcing difficulties. All are extremely secure, and allow windows to be open at night without any worry that anyone could enter uninvited.

They provide an enormous amount of solar gain to the house, since one side, at the front on my house, from top to bottom, is window. This requires a certain amount of active management of blinds-raising them to allow the solar heat in, and closing them or opening windows when the house is warm enough. I did have to accept the various issues that arise from having windows that have to be shipped from another country (in this case Austria). For example, my builder broke the window of a big sliding terrace door and it took three months to get a replacement, during which time the door could not be opened. Overall though, I am very happy with the choice.

Kitchen and appliances

Throughout the house I have few electrical goods. I have a TV with an internal DVD/CD player; a small CD player/radio; a microwave; undercounter fridge with a 6 inch freezer compartment; an electric cooker; a toaster; a soup blender; a small dish washer I use when I have guests over; a washing machine; an air flow heat exchanger; a computer and internet access. All of my electrical appliances are AAA+, with the majority used relatively infrequently because I live alone, work too much, have to travel and do not cook very much.

Bathrooms and water heating

I have always been clear that there are certain comforts in life which are really important to me – big, comfortable bed, bath and sofa; and a spacious shower which has easy on-off / temperature controls. As a result, I have bought a big bath I rarely use but when I do I am reminded of how incredibly comfortable it is. I also wanted to have an en suited bathroom so that when I have guests I can have a separate space to myself. The main bathroom and the en suite have been both the carbon and financial extravagance of the house, although in practical terms, the en suite is a very useful area, particularly in winter, where I can dry clothes. In effect it is used as a laundry room, a significant facilitator in being able to avoid using a tumble dryer in the months of the year when laundry cannot be hung outside on a washing line. If I undertake several washes on the same day, as I often do because of my lifestyle, I could turn on the air exchanger to get rid of excess moisture. However, I have never found it to be a problem.

My water heating is via a solar water heater. Apart from a few hours over the Christmas holiday and January, all my water heating has occurred though the SWH. The control panel shows the water temperature but even so to begin with I had to force myself not to turn on my natural gas water boiler because I could not believe that my water could be hot enough given a cold, grey day outside.

Low energy bills and comfortable living

I have been surprised at just how low my energy bills have become; how successful and efficient the solar water heating is; and how comfortable an energy efficient house is to live in. It is very difficult to compare bills pre- and post- refurbishment because of the poor condition of the house when I bought it and because my house is now bigger. Gas bills were around £60/month and electricity bills were about £50/month. Between August 2010 when I moved into the finished house and spring 2011, I have used gas for water heating for only a few hours in December and January and for space heating for a few more hours than that from December 2010 to March 2011. Otherwise, the solar water heater was enough to provide all my water needs and space heating was most often from a wood burner using compressed sawdust briquettes. The central heating has only been used on cold winter nights when, arriving home late, it has seemed too much effort to put on the wood burning stove. The house does not seem to ever get below about 62–3 degrees F, even on the coldest of days outside and with my blinds down during the day. My monthly gas bills have not been finalised yet. I have gone down from £60 a month to £20 a month but expect to go down again once it becomes clear how little I use. Similarly with my electric bill (Good Energy's Renewable Energy) I currently pay £8 a month. Overall, it seems as though I have cut my bills by around £100 a month.

My low energy bills have led me to the decision not to install photovoltaics (PV) on my roof. I always intended to; indeed the required electrics have been installed within the house so effectively all I have to do is connect up the PV. However, with such low electricity bills; and given my electricity is renewable, from Good Energy, I have decided that it is not worth my while to do so. I understand I could get a feed-in tariff (FIT), thereby making an economic return. I find myself not wanting to do that, I think, because I understand the distributional and equity issues. I somehow feel that I do not want to be responsible for someone I do not know, probably with less money than I have, paying their very small 'socialised' bit of my FIT.

Hindsight and changes

Decisions relating to the size of the refurbished property were based on my house-buying history. This is my third house, with my previous two houses both needing to be refurbished, although not to the extent that this one has been. Having lived through bit-by-bit improvements of my last properties, I was keen to have the refurbishment of my coastguard cottage undertaken in one go. In addition, I had moved from Oxford, which was a two up two down property and just slightly too small for my needs. I used my second bedroom for storage and my living room for working in, which meant it was cramped when guests would stay. I expected, given the Falmouth house is right on the coast, a lot of guests and so decided that an en-suite bathroom would allow me my own space, which I could shut the door on. It was for these reasons that I opted to have two bedrooms upstairs with a third bedroom turned into the en suite shower room and a storage room– just that bit extra than the Oxford house.

In hindsight, there are changes that I would make on the back of my experience. The house now has upstairs – two bedrooms; a main bathroom with a bath, a shower and a lavatory; an ensuite bathroom with a shower and a lavatory; a storage room – and downstairs: two spacious main rooms, one with the kitchen at one end of it; a little entrance space and a downstairs lavatory. This has left me with a house with about one-third more space, with the two rooms downstairs being 30 by 14 feet and 36 by 15 feet. It is not big but if I was to do it over again, I would aim for a smaller house. Most likely, I would take it back to the original small house and place an atrium at the front with one bathroom upstairs and a downstairs lavatory. After all, the overall intention was to end up with a low-carbon house and fundamentally, this alternative proposal would mean that I would not have had to knock anything down or build anything up. I do get enormous pleasure from my house but I think I would get an equal amount of pleasure, and be able to live equally as happily if it was that little bit smaller, with less embodied energy within the refurbishment.

Having lived in the house through a cold winter, I would not have central heating and radiators again, because I so rarely use them. An electricity boost for water and space heating would be adequate for the rare occasions that I need something more. Moreover, I probably would not put in an air exchanger. There is so little air turnover in the house, cooking smells and so on can linger. The idea is that the air exchanger allows that air turn over without the drastic drop in temperature that would occur if a window were to be opened. I have used it in this way a few times but in general I am happy to open the windows. Possibly with a bigger family house an air exchanger may be vital but I have not found that.

Taking on the challenge

All sorts of stresses and strains come into having one's own house refurbished. I had to move out into rented accommodation for nearly a year and the effort of combining a busy working life with on-site visits and on-time decision-making was immense. Much of the very technical meetings between the builder and project manager went straight over my head yet required agreeing to significant expenses. Because the refurbishment was part and parcel of extending my house, and the house was in dreadful condition before, I do not think that turning it in to a very energy efficient house has added much, if any, expense. There are many reasons for this; because timber frame walls are put together so much more quickly than brick houses; because it is such a simple, open design; because I have a small kitchen. Take, for example, my triple glazed windows (including my front and back door). These were no more expensive than double glazed windows from UK household names, although they did cost £18K in total. It has not been the efficiency measures which have cost the most money in the house. The extravagance has been the bathroom furniture and the final size of the house – it could have been smaller (as discussed in Hindsight and Changes). I knew that this refurbishment was to lead to a house which I consider I will be in for the rest of my life and I therefore wanted it to meet very clear requirements. Because I bought it cheaply (because it was in such a bad condition) I knew I could afford to spend up to the amount that it would be worth in a done up state – which I did do. However, I also think, I could have had an equally energy efficient house for less money than I spent.

Lessons learnt

What has become clear to me is that turning an old, leaky, damp house into a warm, dry and comfortable one is not technically complex but it is demanding in that it takes time and it does require thinking about every part of one's lifestyle. For example, whether one wants a big bath; whether one wants to be able to access immediate heating etc. It required external cladding; cavity wall insulation; internal insulation; and wonderful windows. I understand that this requires all the costs to be paid upfront in one go – the windows, the insulation and so on. However, now it is paid for, I have very low energy bills. It has really brought home to me how helpful energy efficiency and SWH can be to the fuel poor. Moreover, I am now extremely sceptical of energy service companies. Once a house is made energy efficient – and as I have said, it's not complicated, energy use plummets. Yes, I use energy, but if all the domestic stock were brought down to my level the domestic energy market would all but disappear. I now consider that this is something very possible to do technically but the barriers are as much with the large energy companies who do not want to see their markets disappearing, as with the energy customers who may not recognise the financial or living benefits.

Lack of incentives to do whole house

At present, the Green Deal is the government's chosen method aimed at banishing the upfront cost and hassle of refurbishing the UK's leaky homes. Although as yet unknown, about £8K will be able to be borrowed which is probably not enough to cover a 'whole house' refurbishment of windows, doors, external and internal insulation. And if it is not enough money, then while it will improve energy efficiency, it will not deliver the 80–90 per cent cuts in energy use that will be required from the sector if targets set in climate change law are to be met.

Socio-technological learning – or lack of?

I think one of the main points I have taken from my experience is that even coming from a fairly well-informed starting point, with a higher than average desire to end up with a low-carbon home and being blessed with the financial capacity to undertake extensive refurbishment, the whole process has often been confusing, sometimes daunting and at times very stressful. I was amazed at how difficult it was to obtain information, guidance and informed advice on building a low-carbon home.

The field of low-carbon building design is relatively new compared to that of traditional building sector, leaving little in the way of established guidance around the optimum technologies, materials or combinations of the two and it is therefore difficult to make choices. If this is lacking, there is even less in the way of linking the application of structures and technologies to individuals and households and the varying lifestyles and needs.

Conclusion

In terms of the house I have ended up with, I am very happy. I love living in it and I am very glad that I undertook the refurbishment. I think a big part of its success is due to the time I spent thinking about the way I want to live and the things in life that are important to me. It is very quiet because of the triple glazing; it is very light and sunny because of the big windows on the south side and the passive lighting; I have comfortable chairs in positions with beautiful views that allow me to be 'still'; and on a day to day basis I love the feeling that I use very little energy, despite having such a practical and comfortable house.

9
Land Use Scenario 2050

Grace Crabb and Adam Thorogood

'The landscape is not just a supply depot but is also the oikos – the home – in which we must live.'

Eugene Odum

We know that climate change brings with it a complex mixture of weather conditions: higher temperatures, increased precipitation, more intense wind speeds, more likelihood of drought, and greater likelihood of flooding. Yet these are so varied from region to region that it is hard to say with certainty what will happen where and when and with what severity. However, we can be sure that agriculture and food production will be threatened by adverse conditions in many parts of the globe between now and 2050. In fact, climate change is already beginning to bite, and close to home. The heat wave of 2003 saw crop failure across Mediterranean regions and cost the agricultural sector in the European Union €13 billion (Paustian, 2006; Easterling et al., 2007), not to mention the loss of human life. As we move towards the middle of the twenty-first century, these seemingly infrequent extremes will begin to happen more regularly, putting pressure on the global food system.

People living at lower latitudes will be hit hardest by climate impacts, another challenge to contend with for those already stricken with poverty. Changes in the climate system, so central to rain-fed agriculture, will hit hardest in regions least able to respond and in places where food production is already marginal. Mid to high latitudes could benefit from small temperature rises and crop types currently confined to southerly climes will have a larger, more northerly range such as soya or maize (Gornall et al., 2010). Longer planting windows and more favourable growing conditions are other changes that could make agriculture more productive in northern latitudes. However, any benefit brought about by increased temperatures could be offset by changes in levels of precipitation; too little in summer leading to drought and too much in winter leading to flooding and the washing away of precious topsoil. Crops could also benefit from greater levels of carbon dioxide in the atmosphere, boosting photosynthetic activity, but again, plants experiencing heat and moisture stress will be unable to take advantage of such conditions. Climate change impacts on the food system are complex and hard to predict. The manifestations of climate change can synergise with other agricultural pressure points such as water scarcity or increased ozone levels to create huge shocks to the food system.

With global population set to increase to 9.1 billion by 2050 (IPCC, 2007) more land will have to be brought into agricultural production. The Food and Agriculture Organisation (FAO) of the United Nations (UN) anticipate an 80 per cent increase in food production by 2050 (FAO, 2007). This, they say, will be the result of improved agricultural technology and land expansion, but with marginal land already threatened by climate impacts, this could be difficult. Also, land currently not productive in the agricultural sense is usually forest or upland, sites that are vital for biodiversity, carbon sequestration and other ecosystem goods and services (Rounsevell, 2006).

The UK context

The British landscape has the capacity to provide a healthy supply of food for its people, a rich habitat for its wildlife and a means of reducing concentrations of carbon dioxide in the atmosphere, despite a projected population increase of 15 million by 2050 (Johnson, 2010). However, conventional agriculture is progressing in a direction that reduces the capacity for these services; perpetuating a system of land management that is detrimental to natural systems, animals and people.

The greenhouse gas emissions from current farming systems are considerable, with high stocking densities, systemic use of fossil fuels and agricultural chemicals. Emissions of methane from livestock and the release of soil carbon and nitrous oxide through crop production and land use change are the largest contributors. Modern agricultural practices have resulted in a polluted landscape that is responsible for nine per cent of the UK's greenhouse gas emissions (DEFRA, 2011). Peak oil and peak phosphate challenges will force the sector to find alternative mechanisms for providing land use services.

Current agricultural policy has brought about large monocultures dependant on fossil fuels, reliant on chemical fertilisers and pesticides, and supporting low levels of biodiversity. This shift has created a fragmented landscape with isolated pockets of semi-natural habitat. Soil productivity is also maintained by application of livestock derived manures, a net source of the potent greenhouse gas, nitrous oxide.

Today only a small per cent of the population are employed in land-based livelihoods. Small family run farms have given way to large industrial farm units with more machinery and fewer labourers, with a push towards increased imports and a reduction of self-reliant food production. The UK population is concentrated in urban centres with little involvement or understanding of the food system and farming. A 'just in time' supermarket delivery system distributes highly processed convenience foods, resulting in nutritionally poor diets.

European government policies have created a subsidised farming system that does not recognise locally sustainable farming practises and overlooks environmental and cultural degradation. This, combined with the free market and globalisation policies, has allowed an influx of cheap imported goods that undermine local markets and carry a high carbon cost due to unsustainable production, transport and refrigeration.

Our current diet is high in white sugars and fats, responsible for a third of heart disease and a quarter of cancers nationally (Tudge, 2007), and an epidemic of obesity. Under current trends, 60 per cent of men and 50 per cent of women could be clinically obese by 2050 (Vandenbroeck, 2007).

The 2050 scenario in this chapter depicts the phasing out of conventional agriculture and the adoption of a sustainable approach to land management that is based on sound ecological, social and moral frameworks. This new system relies on a combination of science and local farming expertise which will allow maximum food production from the UK's 18.5 million hectares of productive land, while protecting natural assets and the ecosystems goods and services provided by areas such as the UK's peat-rich uplands. This national self-reliance in food will avoid the impact of carbon costs associated with imports, and create a secure food supply in an uncertain future where the global food supply could be impeded by climate disruption and subsequent loss of yields abroad.

We anticipate a land use sector that

- Builds soil structure instead of destroying it.
- Keeps cycles of 'waste' products on-farm for use as raw materials in other processes.

- Enriches biodiversity while maintaining optimum levels of production of food and fibre.
- Manages the UK landmass as a net store of carbon.
- Diversifies and provides many more goods for local and national supply.
- Embraces the generation of renewable energy for local district supply and for a decentralised national grid.
- Employs many more people, with young people aspiring to careers in the sector.
- Communicates and cooperates with government and academia to implement policies and research findings.
- Utilises developments in technologies such as GIS mapping and remote sensing to create a more effective landscape scale agriculture.
- Delivers a healthier, low-carbon diet to the nation's tables.
- Involves more people in rural jobs and a great rural resettlement where people return to work and live in the countryside.
- Involves more community participation in the food system through community supported agriculture and more people growing their own food.

Carbon sequestration and carbon trading

Land management will play a key role in the decarbonisation of UK society, not only through reducing emissions by the sustainable production of food and biomass but also through mitigating the emissions of other sectors. This will be achieved by mopping up carbon emissions through the absorption and storage of carbon in well-managed ecosystems.

The land in this 2050 scenario will be managed to absorb carbon dioxide from the atmosphere and to protect and enhance the natural sinks that already contain large amounts of carbon, such as upland peat ecosystems and existing forest soils. The large-scale afforestation of unimproved grassland and marginal land will create new woodland ecosystems that will absorb and store carbon in forest soils and above ground biomass (Brainard, 2003). This potential carbon store will be increased by woodland management practices such as continuous cover silviculture, producing good-quality timber for building materials. This flow of timber into the built environment will increase the sequestration potential of UK forests by locking woody biomass carbon away in buildings and providing space within the forest for regeneration of trees (Milne and Brown, 1997). Integrated farm systems such as agroforestry will produce food yields such as nuts, fruit and animal products while maintaining intact forest ecosystems, thus increasing carbon sequestration. Such multifunctional systems providing a multitude of products, goods and services are typical of 2050 farming in the UK.

Soil is a very useful tool for carbon storage. Good management can increase its capacity but UK soils may reach saturation by 2050. Soil carbon will have to be maintained and monitored under a changing climate to safeguard carbon stores. Harvesting and use of wood products will provide increased pathways for carbon storage beyond that maintained in soils and above ground biomass. As mentioned, it is vitally important that land that is currently not tilled remains a carbon store through non-tillage systems.

England alone stores 584 million tonnes of carbon in its peat uplands; it is the country's largest ecosystem carbon store (Natural England, 2010). Through good practice such as drainage blocking, re-wetting and the re-vegetation of exposed peat, this amount will be increased. Low level grazing through deer management will provide a low-carbon food source and prevention of scrub regeneration, which could damage peat carbon stores (Dawson and Smith, 2007).

Farmers and land managers in 2050 will be adept at managing their land for optimum carbon sequestration. Carbon measurement and monitoring will be the norm on farms, with technological advances making this an easier and cheaper task. Carbon management will be a lucrative business, bringing new money into rural areas.

It is difficult to predict decisions about carbon valuation and trading. So far, carbon trading has yet to deliver any credible reduction in greenhouse gas emissions, and the ongoing debate about the details of the mechanism has been a grave distraction from the growing threat of climate change. The use of a market based mechanism to trade carbon is seen as highly controversial by some circles (Kill et al., 2010), who see it as way of the financial sector making large sums of money from a tradable commodity without delivering climate change mitigation. For this scenario we predict that there will be some kind of financial incentive for choosing sustainable land use practices that sequester carbon and that this is considered to be a central aim of farming and forestry alongside food and fibre production (Ostle, 2009).

The landscape of the future

By 2050, efforts made to slow and reduce climatic disruption and biodiversity loss, coupled with new environmental conditions will change the appearance of the British landscape. Low-carbon farming and forestry activities that sequester carbon while protecting wildlife and producing food and materials will be the norm (Liddon, 2009). Farms will generally be more diverse than present, and the landscape more mosaic in appearance, with diverse cropping and livestock management suited to specific regions. The landscape is covered with more trees both in forest ecosystems and integrated into farming systems, helping to prevent large scale soil loss and providing shelter from strong winds. Large interconnected hedgerows (or shaws) divide grazed and arable fields, providing wildlife habitat corridors and a wide variety of edible wild and domesticated fruits and nuts, and many wood based products. Farm ponds and lakes are a much more common feature of the British landscape, capturing water in the winter, in particular in the drought prone South. They also provide a source of fish protein for local markets, algae for biomass and duckweed for on-farm chicken feed.

Crop types will change by 2050, with a greater variety of species being suitable to the UK climate. This will favour agro-forestry systems using grapes and other less frost tolerant fruit and nut trees. Perennial crops for biogas production are a feature of the 2050 landscape, forming a small component of most farm acreage. Rather than the rolling improved pastures that have become familiar in many parts of Britain, grazing fields form a smaller piece of the landscape puzzle. Ruminants are fewer in number due to the emissions related to their life cycle, but those that are kept provide a vital function in improving soil carbon and fertility. Grazing is maintained in organic rotation with soil improving techniques that build organic matter and fertility. By 2050 there will have been a 90 per cent reduction in the numbers of grazing ruminants in Britain. This will not only reduce direct methane emissions from the gut of the animals, but also reduce indirect nitrous oxide (N_2O) emissions from manure spreading (Eckard, 2010). We expect to see only ten per cent of current beef numbers, 20 per cent of sheep and 20 per cent of dairy cattle farmed. This frees up nine million hectares of previous grazing land where nutritious, edible crops for direct consumption, biomass, feed-stocks and silage for anaerobic digesters can be grown (Kemp and Wexler, 2010). By 2050 many farms integrate grazing, agro-foresty, arable crops, perennial energy crops and woodland systems to form a complex and diverse landscape (OST, 2010).

Biodiversity

By 2050 the current trend of biodiversity loss is reversed, and the landscape plays host to rich and resilient ecosystems. This change is made possible by a u-turn in current political thinking, initiated by successful conventions on biodiversity conservation. The protection of species is brought about by policy and praxis united by the collaboration of scientists and agricultural practitioners. Despite the widely unknown nature of ecosystems interactions, the law begins to recognise the potentially catastrophic risk to agricultural, natural and social wealth through the collapse of species and whole ecosystems. Landscape-scale intervention follows, with the careful reconstruction of key ecosystems by the farming community, government and non-governmental organisations. These groups work together to share knowledge and restore the landscape to one that is both rich in biodiversity and in food supply.

Farms cooperate to plant wide hedgerow corridors that are thinned for products. These shaws are planted with locally appropriate species that support a complex and unique shrub layer, ground flora and subsequent fauna. Organically managed crops and pastures encourage wild flowers and insects to thrive in a balanced ecosystem, and as a consequence support the rest of the food chain. Managing rough pasture for silage, biomass and grazing encourages rich meadows with regionally distinct flora and fauna. Woodlands are managed with mixes of locally native tree species and non-native timber trees, and ground flora is protected to provide habitat for woodland species. By increasing the size and interconnectedness of certain habitat types, the worst effects of climate change will be buffered. Species will be able to migrate north in warmer periods and find food sources in adjacent land. The study of localised migration will allow landowners to better understand planting mixes and habitat management necessary for their region. Landowners and managers monitor population densities of key indicator species in order to understand changes to the health of any particular system. These changes will require a new approach to landscape management where farming and conservation are no longer distinct. It is critical that farmers and foresters become conservationists on the one hand and the providers of food and raw materials on the other. Although this is a lofty aim it is not impossible to imagine, and will be hugely dependent on appropriate subsidy schemes and the development of new supply chains.

Financial support

This transformation will be supported by carefully constructed agri-environment schemes that preserve our natural and cultural heritage while providing food for Britain. Farm carbon accounting is commonplace and incentives are provided for carbon savings. Current agri-environment schemes reflect a valid concern for ecosystem health, biodiversity and carbon storage, but all too often at the expense of food production. By 2050, farm payments both protect the environment *and* produce high yields for British consumption in integrated models. This is achieved through multi-farm landscape scale planning of agricultural, wooded and natural land. Numerous farms will work together on schemes such as water catchments and wildlife corridors, while coordinating crop plans and fertility measures to ensure maximum productivity.

The change in farming livelihoods

As opposed to the increasingly isolated livelihoods of current farmers, by 2050 we will see highly cooperative planning at a regional scale. Farmers will join together sharing knowledge, data and infrastructure, as food production is closely coordinated across farms. Produce is pooled to create viable local markets providing a rich range of services and products. Market halls within towns will enable farmers to sell their products directly to consumers through the farming cooperatives. This new approach is facilitated by Rural Environment Forums where farming cooperatives communicate directly with local and national specialists, scientists, policy makers, ecologists from non-governmental organisations and other land managers.

Case studies

The farm will become a multifunctional enterprise where food is grown, natural materials, timber and biofuels are produced, water is conserved, carbon is sequestered and biodiversity is enhanced. Houses and infrastructure for a growing population will be sensitively located so as not to impinge on these achievements. This will require significant shifts in current farming practice. The most dramatic of these may arguably be the large reduction of livestock, particularly ruminants such as sheep and cows. The driving force behind changes in agricultural practice is the vital necessity to reduce emissions and to absorb the excess emissions of other sectors.

South East Farm 2050

The typical southern lowland intensive farm is projected to go through radical changes to its practises by 2050. High output forage systems (improved grazing) are destined to undergo some of the most drastic alterations. The following describes the transformation of one such 90-hectare beef and sheep lowland farm.

The Suffolk sheep and dairy farm sees decreasing stocking rates throughout the early twenty-first century. Sheep are entirely removed from the system and the number of cattle reduced by 80 per cent to a head count of 20 animals. Approximately half of the current improved pasture (37 hectares) is returned to a traditional rotational ley system where beef cattle are rotated with field beans, cereals and clover green manures (see Figure 9.1). This diversification provides a wider range of products for direct human consumption, and a more efficient system of nutrient recycling. By 2050 the farm uses only on-site generated fertility and maintains its fertility from its own from legumes, manure and composted effluent from the biogas generator. This compares with ammonia nitrate fertiliser use in 2010, which had embodied carbon costs of 6.5 tonnes per year.

An additional four hectares of the farm is dedicated to walnut crops, requiring 100 per cent less water than previous improved grassland, and no imported nitrogen fertiliser (Stavins and Richards, 2005). These agroforestry systems are planted as alleys of trees with crops in between. These crops are on a rotation with animals, and provide fodder for the cattle but also arable crops for human consumption. Pigs, geese and chickens are turned out among the trees, and provide small scale meat and egg production while improving soil fertility and crop production. The meat market has contracted in the past 40 years, with sales reaching a peak during festive periods. Small-scale horticulture is employed on the farm, with high value production of mushrooms (see Table 9.1).

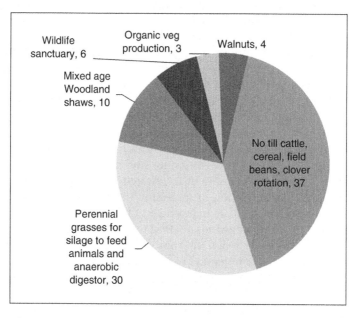

Figure 9.1 Hectares required for each system

Table 9.1   Annual quantity of product range from the South-East lowland farm

Commodity	Average quantity exported per year	Average quantity used on farm
Pigs	140 pigs	2 pigs
Geese and chickens	50 geese/chickens	5 geese (staff)
Eggs	15000 eggs	2000 (staff)
Walnuts	24 tonnes	–
Silage	1200 raw tones (307,200 KWh?)	4 tonnes (fodder)
Cereals	107 tonnes	4 tonnes (fodder)
Beans	148 tonnes	4 tonnes (fodder)
Cattle	10 cows	–
Woodland craft products	Variable	–
Mushrooms	2 tonnes	0.08 tonnes staff
Organic veg. mixed	100 tonnes	3 tonnes (staff annual)

Berkshire pigs are selected for their long hind gut that can draw more nutrients from cellulose. Six family herds, ten sows in each, are rotated among the other sites on the farm. The pigs are supplemented with small quantities of site-grown grain, and swill from municipal pasteurised compost that is not used in the anaerobic digester (see below). The pigs are also brought into the yard on certain rotas in order to capture manure for fertiliser application.

Previous land used to grow fodder crops is now freed up for the growing of wheat that is directly consumed by humans. Over the course of 40 years, this farm has worked closely with the scientific community to develop locally evolved strains of organic wheat with longer stalks that do not compromise the size of the head. This has allowed for an organic form of production where weeds do not smother the variety. This government-led drive for investment in research across Britain has led to the breeding and cultivation of successful, high yielding varieties that require lower inputs, and are independent of chemical fertilisers.

Carbon equivalent emissions of five tonnes were generated upon the initial land conversion from improved pasture to rotational leas. This efflux has been compensated for by the increase in sequestration through woody biomass and soil carbon (see Table 9.2). The farm now has a 50 Kilowatt anaerobic digestor, using silage grown on the perennial grasslands. These grass species have been selected as low-input varieties, reducing related emissions from fertiliser production. Many other farms

Table 9.2 Land practice carbon flux at South-East lowland farm. Influx (+) and efflux (–) in 2050

Product	Carbon/t
Walnuts	19.2 tonnes (+)
Hedgerow shaws	39 tonnes (+)
No till rotation	22 tonnes (+)
Grassland	30 tonnes (+)
Cattle	92 tonnes (–)
	Total sequestered: 28.2 t/C

in the region have also de-stocked and replaced animals with perennial grasses. This farm has an aerobic digester that processes grass grown on permanent pasture and within the ley system. It is supplemented by sewage collected from a local sewage works. Improvements of sewage collection have prevented heavy metal contamination. The slurry released from the digester is then applied to the land.

Water conservation measures are practised at the farm, reducing groundwater abstraction. Crops are irrigated at night and mulch is used to prevent evaporation on arable crops. New irrigation methods such as flood and drip irrigation are practised, and water is stored in catchment scale reservoirs around the farm.

This new style of farming was made possible by the initiation of cooperative farming networks in the area. This farm is a member of a regional cooperative that exchanges knowledge, current practise, and crop information. This collaboration also allows for the production of what is needed within the region. A larger national network allows the trade of staple crops across Britain to where they are required. All products are bought and sold through a farmers' market system where the general public, service industries and supermarkets can purchase the goods from their locality.

This farm is also subsidised by a robust agri-environment scheme which, through rural forums and government strategy, has created a long-term management plan for the farm, detailing future carbon sinks and sources, water conservation and biodiversity enhancing strategies. The five farms adjacent to this example are together managing the range of the Tree Sparrow, *Passer montanus*, which uses a stretch of wide hedgerow to move between feeding and breeding sites. This hedgerow is also managed for firewood on a rotation that has been agreed in the scheme. All farms have adopted these locally appropriate schemes encompassed within national objectives.

The workforce on the farm has increased from five in 2010, to 30 full-time equivalent staff composed of permanent and seasonal workers offering a range of crafts. Products are developed from the on-site materials such as native tree species for greenwood crafts. Animals are slaughtered on-site and value is added through the curing of bacon. Staff are fed once a day from products grown on-site. All activities occur within the farm-building complex, and operations are open to visitors. Most craftsmen run an apprenticeship scheme and farmers provide educational knowledge shares in order to develop people's agricultural knowledge.

Cambrian Farm 2050

In the future, the upland farm becomes increasingly valued for the provision of ecosystems goods and services. Intensive sheep farming gives way to diversified land management, producing a range of products and services including small amounts of high value meat, timber, wood fuel, short-rotation willow coppice (for biomass energy), honey, peat land carbon sequestration, deer management, water quality and flood mitigation. This shift in focus also sees a revival of upland rural communities and enterprises. The renaissance of upland farming sees more young people staying on family farms or starting their own rural businesses (CRC, 2010). An influx of workers sees the composition of upland communities change; this re-ruralisation brings new pressures to rural communities but also new opportunities.

The strategic importance of upland management becomes more acute during the first decades of the twenty-first century, as climate change impacts begin to take their toll and force adaptation actions from community level right up to government. We use the example of one upland farm in Mid-Wales to illustrate some of the differences between upland farming now and the future farming of such landscapes.

Cambrian Farm spreads over 30 hectares in Mid-Wales, rising from an altitude of 250 to 470 metres above sea level. In 2050, this farm is a mixture of rough pasture, woodland, woodland pasture and upland peat bog, with linear features such as stone walls and wide hedges dividing up the landscape (see Table 9.3). The land has been farmed for hundreds of years and has seen many changes to farming practice over the years, not all of them easy transitions. The average farm size has reduced in the uplands, creating smaller units that are more easily managed by the farming family and a team of local employees. Clusters of small farms form operatives sharing tools, equipment, labour and skills. Cambrian Farm is part of the local *Hill Farm Co-op*, a group of five upland farms of roughly the same size that total 200 hectares.

Table 9.3 Annual quantity of product range from the Cambrian farm

Commodity	Average quantity exported	Average quantity used on farm
Firewood	60 m³	
Firewood from coppice		12.6 m³
Timber logs	80 m³	
Coppice biomass		20 oven dry tonnes
Meat (venison)	300 kg	
Meat (lamb)	530 kg	100 kg
Wool	340 kg	
Lamb skins	35 skins	
Honey (from 40 hives)	915 kg	5 kg

Smaller farm units allow for a focused stewardship of local ecosystems; the family know their land intimately and can monitor changes in biodiversity and species composition as well as assessing soil carbon flux and water through-flow. The farmer can also monitor changes to the landscape and local weather patterns due to climate change and adapt his practice to suit. A cooperative approach allows the five farms of the Hill Farm Coop to work together on larger scale landscape projects that aim to produce resilient upland ecosystems, rich in biodiversity and prime for carbon sequestration. For example, peat land and woodland management and renewable energy projects are embarked on as a cooperative, with the five farms pooling resources.

In the twentieth century, Cambrian Farm was a much larger unit of around 300 hectares. Headage payments that rewarded the farmer for production, lead to over-grazing and a concentration on sheep farming above all else. The adverse impact of so many grazing ruminants on the natural and semi-natural vegetation of the Welsh uplands was extreme, leading to degraded ecosystems with a reduced capacity to fulfil vital environmental services and social benefits. The environmental stewardship model of farm payments, beginning in the early twenty-first century, is the model that shapes Cambrian Farm over the years between the present and 2050, away from over grazing with large flocks towards a smaller, more diverse, multifunctional farm. A smaller farm size also reduces the need for large, fuel-hungry machinery. The remaining machines are powered by biodiesel produced by the Hill Farm Coop with the raw material being grown across the five farms and pooled.

The reduction in livestock levels is one the biggest changes at Cambrian Farm in 2050, driven by the re-focusing of farm payments towards stewardship and vital services such as carbon sequestration. Stock levels drop from 2,000 breeding ewes and 200 suckler cows on the larger old farm unit of the past to a drastically reduced herd of around 25 ewes of Beulah Speckle Face, a hardy breed capable of extensive grazing over upland pastures with little supplementary feed (FoE, 2010). This drop in stock levels happened over a number of years through reduced breeding regimes, bringing livestock in line with an optimum number for the rejuvenation of degraded upland landscapes, while maintaining a level of meat production for local consumption and sheep wool for insulation (Taylor et al., 2010). Shearing is one of the seasonal cooperative efforts that bring the five farms of the Hill Farm Co-op together, producing a bulk crop for sale as insulation to the local construction industry. There is a reduction in chemical use to control parasites and a revival of management practices such as rotational grazing and a breeding programme to select for genetic resistance to pests. The pasture has been sown with legumes to improve nitrogen fixation and carbon sequestration as well as having earthworms introduced to improve soil structure and carbon storage. Trees are planted on the steeper sites, the roots help to prevent erosion of topsoils and livestock are grazed in the understory, forming a wood pasture system.

Cambrian Farm's peat uplands have been brought into pristine condition and are closely monitored to safeguard the huge stores of carbon locked up within their structure (see Table 9.4). A river catchment management project involving all five farms in the Hill farm Coop was started with government funds to mitigate lowland flooding. Teams tasked with blocking up peat lands aided farmers to prevent carbon run-off from the uplands and re-wet areas of degraded peat bog.

Table 9.4 Land practice carbon flux at the Cambrian farm. Influx (+) and efflux (−) in 2050

Land use	Carbon/t
Rough pasture	0.25(+)
Wood pasture	0.2(+)
Broadleaf woodland (YC 8)	37.1(+)
Ash coppice	10.42(+)
Short rotation willow coppice	7.8(+)
Peat upland	3.5+
Hedgerow (12 km)	24(+)
Livestock	14.77(−)
Total sequestered	68.5(+)

Reforestation of upland grazing land has also slowed the movement of water down through the catchment. Due to the reforestation of much of the farm, major products in 2050 are timber and firewood, mainly sold into local supply chains. The farm has their own sawmill and adds value to their produce by producing good-quality milled oak for timber frame construction. In comparison to the farm of the past, Cambrian Farm has significantly more woodland cover and the farm workers are skilled woodsmen and foresters with knowledge of sustainable silviculture practices such as continuous cover forestry. Broadleaved species such as oak, beech and chestnut are favoured for their high carbon density.

Short-rotation willow coppice has found new markets in the growing combined-heat and power (CHP) usage by district-scale renewable energy systems. The farm retains some willow coppice for use in their own on-farm CHP unit, providing electricity and heat for use on-site (Tubby and Armstrong, 2002). The excess heat from this unit is used to speed up the seasoning of firewood and timber. The farm and the Hill Farm Coop as a whole have embraced renewable energy. A large wind power scheme now lines the ridge above the farms. Close monitoring of the upland habitat proceeded the commissioning of the scheme that has been put in place with minimum ecological impact. The farms receive a considerable income from government incentives to produce renewable energy for a decentralised grid (Liddon, 2009).

Landscape and society in 2050, feeding people

Such an ambitious transformation of the land use sector will not be possible without the refocusing of national priorities towards an agrarian future where rural livelihoods and communities are of great importance to food supply and home grown production. There will be a large increase in the number of people moving to rural communities from urban centres in order to work in farming and forestry. Although some farm machinery can still be run with on-site produced biofuels, the increase in manual labour compensates for a lack of fossil fuel use, which becomes prohibitively expensive. The influx of people is met with opposition and suspicion at first, but gradually as climate change becomes an acknowledged reality for all sectors of society, a culture of necessity to adaptation and mitigation is established.

Rural livelihoods

Communities work together as efficient units and absorb climate migrants from abroad. Incomers partake in rural tasks as part of a thriving and self-sufficient community. With the recognition of the potential catastrophe of lack of food, and the pressures on food supplies from abroad many workers from other sectors decide to dedicate their lives to feeding the nation. Necessary manual work is complemented by volunteers of all ages who wish to get a foothold in the agricultural sector and those from urban areas who take working holidays to help the cause. Carbon constraints result in less commuting, and therefore local people undertake more semi-skilled and unskilled work.

Housing market pressure increases with greater rural density, encouraging young people to innovate and establish low-cost housing in eco-communities. This is met with initial reluctance by local planning authorities but as the impact of climate change becomes fully understood, planning priorities shift towards adaptation and mitigation favouring sustainable and ecologically sound developments. Lack of rural housing and the high cost of air travel bring families closer together, as they were in the past, however nuclear groups are less common and families form communities where child and elderly care, food supply and transport is shared. This is popular both in new eco-villages and in older housing where gardens are merged, occupancy reallocated and responsibilities shared. Those owning surplus rooms in large homes are encouraged to let any spare accommodation to incomers for the common good. This requires a dramatic shift in the British culture and does not, at first, come easily, but with the realisation of potential threats to food security and habitat loss there comes a change in mentality. Incomers quickly become part of the rural community taking advantage of the increase in leisure activities and communal space available.

Walking and cycling are given much provision in the new 2050 transport plans, and it is rare for towns not to be connected with well-used cycle routes. All villages have a local shops and cafe that provides a meeting place, where locally grown provisions can be bought and sold. Most services work at community level, including shopping, education and leisure.

Urban areas

In cities and towns there is a huge growth in allotments with councils being forced to provide more space for growing due to the high demand. Street and estate scale compost collection schemes provide much needed organic matter for city soils, this coupled with urine collection for fertiliser sees a real boost to nutrient levels. Urban gardens and allotments are managed for carbon sequestration, with gardeners employing biochar to store carbon in city soils. An enlightened approach to urban green space sees the creation of multifunctional sites across towns that provide land for growing, recreation and respite from the stresses of city life. There is a programme of urban tree planting that aids carbon sequestration efforts and reduces the urban heat island effect, many of these trees produce nut and fruit crops. Green corridors are integrated into cities that allow passage from the city centre right out into the countryside allowing increased communication between urban and rural areas. Urban communities who want a closer connection with the production of their food have driven the growth in community-supported agriculture. Farmers, particularly living in the 'food belts' – the market garden zones around major cities, receive an influx of city visitors at key times of year. There is a twinning of city districts and rural food producing areas.

Consumption

The food system is the production of food and goods but their transport, processing, refrigeration, preparation, sale, final consumption and waste. 2050 will be a time of a more enlightened approach to consumption both of food and also of ecosystems goods and services. There will be a more holistic understanding of agriculture's beneficial role. People will pay more for food products in the knowledge that they are being produced by sustainable food systems. Consumers are educated and aware of the necessity to decarbonise the entire food process and add value to the UK landscape.

Trading in community and local markets, and Community Supported Agriculture Schemes (CSAs), encourage a greater awareness and involvement by consumers in the food process. Consumers move away from the passive consumption of high-carbon goods to actively seeking the local and the seasonal, helped by certification schemes. A huge revival of home growing sees city allotments increase in number on urban brown-land sites and city edges. People in 2050 will grow more food themselves with the bulk of their food sourced locally.

Lower meat consumption requires a drastic cultural shift underpinned by a widespread understanding of the necessity to make such a change. It seems radical but it is nothing more than a return to consumption patterns of the past. The high price of meat brought about by the carbon credit system (producing meat will incur a higher carbon cost than other food products) and lower stock levels, will encourage its use as a luxury item eaten during times of celebration. The valuing of meat as a luxury food product will encourage a better approach to animal welfare and a moral dimension to animal husbandry that has been allowed to slip from current intensive systems. Less quantity but much better quality will typify the meat industry, which sells more for celebration than daily consumption.

The typical 2050 diet will be low in carbon-intensive products such as red meat and dairy products and high in vegetables and whole grain foods plus nuts and legumes. A mainly vegetarian diet will be the norm, with high-quality meat enjoyed during celebrations and fish and eggs consumed less frequently. Poultry (specifically eggs) and pigs (being non-ruminants), although fewer in number, still feature in the UK diet of 2050. Fish will be more widely consumed due to increased levels of production in fish farms, but this, as with other meat production, will be highly monitored to establish and continue good welfare and compassionate farming practices.

The shift in diet and food culture from a high meat consumption, together with take-away or processed food with low nutritional value, to a more traditional diet based on locally produced food lead to better public health with a reduction in obesity, heart disease and cancers. This long-term health benefit reinforces the required changes to the farming and food system.

Conclusion

The above scenario may seem unrealistically optimistic under current conditions but drastic change is required. A widespread and deep-seated understanding of the problems caused by climate change and unsustainable land use practices drive these future shifts in land use and the food system. Dramatic weather events shock people into taking action, from grass roots level to corporations to governments. People, threatened by a violent and unpredictable climate system, develop a new, more cooperative attitude. There is no longer any doubt that things have to change and this new resolve drives the move towards a more sustainable form of land management for multiple goods and services. Many of the changes described above are not new; they are derived from sustainable farming and land management systems of the past. Yet coupled with these age-old ways of working the land is an embracing of new technologies that allow for effective and cooperative landscape scale farming. Consumers understand that their purchases have an impact on the climate and are far more conscious of what a sustainable food system actually means. Farming in 2050 is the result of some tough and unsettling decades but by this date, a sustainable system is in place.

References

Brainard, J., A. Lovett and I. Bateman (2003) 'Carbon Sequestration Benefits of Woodland', Centre for Social & Economic Research on the Global Environment, School of Environmental Sciences University of East Anglia.

CRC (2010) 'High Ground, High Potential – a Future for England's Upland Communities', summary report, Commission for Rural Communities.

Dawson, J. and P. Smith (2007) 'Carbon Losses from Soil and its Consequences for Land-Use Management', *Science of the Total Environment* 382: 165–90.

DEFRA (2011) 'Farming Industry must Act to Reduce Greenhouse Gas Emissions', Department of Environment Food and Rural Affairs, 4 April 2011, www.defra. gov.uk/news/2011/04/04/farming-emission.

Easterling, W., P. Aggarwal, P. Batima, K. Brander, L. Erda, S. Howden, A. Kirilenko, J. Morton, J.-F. Soussana, J. Schmidhuber and F. N. Tubiello (2007) 'Food, Fibre and Forest Products', in M. Parry, O. Canziani, J. Palutikof, P. van der Linden and C. Hanson (eds) *Climate Change 2007: Impacts, Adaptation and Vulnerability*. Contribution of Working Group II to the Fourth Assessment Report of the Intergovernmental Panel on Climate Change, Cambridge University Press, 273–313.

Eckard, R. (2010) 'Options for the Abatement of Methane and Nitrous Oxide from Ruminant Production: A Review', *Livestock Science*, 130(1–3): 47–56.

FAO (2007) 'Adaptation to Climate Change in Agriculture, Forestry and Fisheries: Perspective, Framework and Priorities', Rome, Italy, Food and Agriculture Organisation of the United Nations.

FoE (2010) 'Fixing the Food Chain: Benefits of Hill Farming in the Uplands,' Briefing July 2010, London, Friends of the Earth, http://www.foe.co.uk/ resource/briefings/hill_farming_benefits.pdf.

Gornall, J., R. Betts, E. Burke, R. Clark, J. Camp, K. Willett and A. Wiltshire (2010) 'Implications of Climate Change for Agricultural Productivity in the Early Twenty First Century', *Phil. Trans. R. Soc. B* 365; 2973–89.

Kemp, M. and J. Wexler (2010) 'Zero Carbon Britain 2030: A New Energy Strategy', Centre For Alternative Technology.

Johnson, W. (2010) 'UK Population "Largest in Western Europe by 2050"', *The Independent*, 30 July 2010.

Kill, J., S. Ozinga, S. Pavett and R. Wainwright (2010) 'Trading Carbon: How it Works and Why it is Controversial', Moreton in Marsh, Glos, FERN, www.fern. org/tradingcarbon.

Liddon, A. (2009) 'Landmarks for Policy, Rural Economy and Land Use Programme', Briefing Series No. 9, Newcastle University.

Milne, R. and T. Brown (1997) 'Carbon in the Vegetation and Soils of Great Britain', *Journal of Environmental Management* 49: 413–33.

Natural England (2010) 'England's Peatlands – Carbon Storage and Greenhouse Gases', Sheffield, Natural England, www.naturalengland.etraderstores.com/ NaturalEnglandShop/NE257.

OST (2010) 'Foresight Land Use Futures Project 2010', Final Project Report, London, Office of Science and Technology.

Ostle, N. (2009) 'UK Land Use and Soil Carbon Sequestration,' *Land Use Policy* 26S: S274–S283.

Paustian, K., N. Ravindranath and A. van Amstel (2006) 'IPCC Guidelines for National Greenhouse Gas Inventories, Volume 4: Agriculture, Forestry and Other Land Use', Intergovernmental Panel on Climate Change, www.ipcc. ch/meetings/session25/doc4a4b/vol4.pdf.

Rounsevell, M. (2006) 'A Coherent Set of Future Land Use Change Scenarios for Europe, Agriculture', *Ecosystems and Environment* 114: 57–68.

Stavins, R. and K. Richards (2005) 'The Cost of US Forest-Based Carbon Sequestration', Arlington VA, USA, Pew Center on Global Climate Change.

Taylor, R., A. Jones and G. Edwards-Jones (2010) 'Measuring Holistic Carbon Footprints for Lamb and Beef Farms in the Cambrian Mountains Initiative', Policy Research Report No. 10/8. Countryside Council for Wales, www. cumbriacommoners.org.uk/files/measuring_holistic_carbon_footprints_for_ lamb_and_beef_fa_pdf.pdf.

Tubby, I. and A. Armstrong (2002) 'Establishment and Management of Short Rotation Coppice', Edinburgh, Forestry Commission, http://www.forestry.gov. uk/pdf/fcpn7.pdf/$FILE/fcpn7.pdf.

Tudge, C. (2007) *Feeding People is Easy*, Italy: Pari Publishing.

Vandenbroeck, P., J. Goossens and M. Clemens (2007) 'Foresight – Tackling Obesities: Future Choices – Building the Obesity System Map', London, Department for Innovation, Universities and Skills.

10
Zero-Carbon Britain: Time to say 'we will'

Tanya Hawkes

The strap line of *zerocarbonbritain2030* (ZCB, 2010) reads: 'the science says we must, the technology says we can. Time to say we will.' This chapter explores what kind of society would willingly say 'we will' to rapid decarbonisation. What kind of national government would be required to enable this vital transition, its policies, ethics and values? What would this society contribute to the international climate change process? The chapter argues the case for democracy and for a renewal of the welfare state as the ethical option to a zerocarbon transition, within an international arena that places environmental justice at its heart.

Zero-carbon Britain – an idea comes to life

In 2006, a group of students and staff at the Centre for Alternative Technology (CAT) gathered together to do a fun project: finding out how quickly and feasibly Britain could reduce its carbon emissions to as low a figure as possible. They had no funding, just an abundance of enthusiasm. The idea spread and within months scores of staff and students at CAT were meeting in the evenings and lunch breaks. Groups formed that examined different sectors such as buildings, transport and energy.

During the last few weeks of the project, through a wiki system, the participants updated the last few bits of data. The result – that it would take 20 years to descend to zero-carbon emissions, by powering Britain entirely on renewable – surprised everyone. The small, underfunded research project became a phenomenon. The All Party Parliamentary Group on Climate Change requested CAT launch the project in parliament and the huge group of staff and students from CAT who had stayed up endless nights collecting data were astonished and delighted that *zerocarbonbritain* (ZCB) – as it was retrospectively named – was being taken so seriously (ZCB, 2007).

The Centre for Alternative Technology has a long history of testing out radical technologies and ideas. Its 140 staff still reach decisions by consensus and the thousand Masters students, many of whom are also staff, mix in. It's a true learning organisation. There is a constant experiment being conducted: in the co-operative management system, the micro grid energy generation, the water and sanitation systems and the research. CAT lacks funds and resources but it never lacks motivation (CAT, 2010).

As with any research the first edition of ZCB raised several questions. Its critics were concerned that it had not thoroughly addressed the issues of land-use and agriculture, or focused enough on policy and behaviour change, so when the second phase of ZCB – *Zerocarbonbritain2030* – began, it was with appropriate funding and resources. Thirteen universities, 12 research bodies and NGO's and eight companies pooled nearly 120 experts to work on the project for a year. Nine areas were researched: climate science, energy, the built environment, transport, land-use and agriculture, renewables, policy and economics, employment and behaviour change. The co-ordinators hosted a research week for each area, followed by a public seminar to scrutinise and test the findings for each. Some were much more contentious than others.

Meat and flying

The first transport research seminar began with a heated debate over aviation. One participant stated from the outset that ZCB needed to include a high level of aviation allowance: 'If aviation isn't allowed, no policy maker will take it seriously.' The almost unanimous response was that ZCB was not interested in what was political acceptable, only what was technically possible. The task of changing the high-carbon behaviour and lifestyle choices of the electorate was beyond our remit. We would explore the research and some of the challenges associated with policy and behaviour, but to worry about the politically viable nature of our findings before we even began would weaken our research. The recommendations for policy and behaviour change interventions would need to fit around the science and technology, not the other way round. There were to be no political boundaries.

A picture of a zero-carbon society: The unimaginable?

One of the slides for the ZCB2030 presentation says: 'If we can't imagine the future, how can we create it?' A picture of the sci-fi world follows that many of us are familiar with from films and books. It looks like this: our societies, particularly the westernised section of the long industrialised countries will keep on progressing. Technology will improve as far as the learning curve allows and we will invent new and convenient ways to travel and entertain ourselves. Vehicles will fly through the air and personal communication systems will get smaller and sleeker. We will be wearing intelligent clothes and space travel will be commonplace. We grew up with images from the films *Star Wars* and *The Fifth Element*. The flip sides of these are the apocalyptic stories. The film dystopia of *Mad Max* and *Blade Runner* to the hopelessness of Cormac McCarthy's book *The Road*. Robert Mckee tells us that stories are human 'equipment for living'. He goes on to say 'Our appetite for story is a reflection of the profound human need to grasp the patterns of living, not merely as an intellectual exercise, but within a very personal, emotional experience'(Mckee, 1997: 12). He argues that we make our lives into stories; the past, present and future, to give ourselves meaning. How are we supposed to construct these new narratives? In a society where much of the economic progress that we consider exciting: cars, air travel, entertainment and information technology involves carbon-intensive technology. For many the low-carbon alternatives, like cycling and re-using consumer items, are considered boring or regressive. The disparaging term 'less developed' is the used for nations who are not part of the high GDP club.

While it may be possible to decouple progress from economies based on fossil-fuels in the future, it's hard to imagine how the transition might look. Although there are many admirable texts that outline the possibilities such as *Beyond Growth* by Herman Daly (1996) and *Prosperity without Growth* by Tim Jackson (2009), there are few that really illustrate the transition and its impacts on people's lives. Perhaps *Energy and the Industrial Revolution* by Wrigley (2011), gives some of the best illustrations as it links the reader to a time before fossil fuels, when the pace of growth was constrained by the available energy from land-use. The need for food kept forestation for fuel in a sustainable balance. It's a picture of the past, not the future, but it brings alive an energy system that is invisible to most people. Renewable technology allows for greater growth, but perhaps not with the acceleration of a fossil fuel economy.

Picturing policy

Even harder to imagine than future energy systems, land-use and agriculture and transport are the more abstract areas of policy and behaviour change. Policy and economics is not usually considered the stuff of stories. It's hard to read the climate change white paper – the *Low Carbon Transition Plan* (DECC, 2009) – and allow one's imagination to shape it. The dry and exclusive language prevents it. And yet policy changes people's lives, and redirects the course of society. Policy is a world of people and people lives become stories.

There is a joke in NGO and civil service circles that if you want something to happen without anyone noticing or if you do not want anyone to read about something, then produce a document on it. So it seems right to try and bring some of this policy and legislation to life, or to imagine the future as shaped by international and national environmental policy. The Committee on Climate Change, under the Climate Change Act legislates for five year, decreasing carbon budgets. To enable that to actually happen takes lateral planning. Planning is arguably just well informed imagination. When Anthony Giddens (2009) talks about the need for a return to planning and backcasting, he is espousing a method that requires foresight and risk.

The key events we read in history are shaped by policy and legislation, from the enclosure sparked rebellions to the struggle for suffrage. Decisions made in UK parliament cause large scale changes to other countries. In the small print of long documents and complex graphs are the details that affect the lives of millions.

The Foresight reports from the Office of Science and Technology (2006) reflect this need to visualisation of the future, combining predictions of intelligent infrastructure with the constraints that global warming and climate change. Four scenarios are mapped to show possible effects of different societal choices in transport, energy and IT. 'Perpetual Motion' is an urban environment, fast paced and running on clean energy. In 'Urban Colonies' there is a dense but more localised city scenario, where the impacts of climate change have forced more localised economies. In 'Tribal Trading', after significant energy shocks that results in the dismantling of the national grid and the abandoning of cities, there is the creation of a rural self-sufficient society. 'Good Intentions' describes a society, which tries to reduce social and economic inequality, through personal carbon taxation systems.

What makes this document stand out is the attempt to put the backcasting style scenarios into a visual and creative format. The data is sandwiched between artists' impressions and fictional interpretations of how these scenarios would impact on families and individuals, transforming the language of policy into stories and images.

ZCB2030 imagining the impossible

The pages of ZCB take the reader through a journey from international policy frameworks that combine environmental justice and carbon capping, right through to the number of houses that would need retro fitting with insulation. The detail is painstaking and comes from many years of hands on expertise, combined with meticulous research.

It recommends that *buildings* undergo a massive refurbishment programme, with a large amount of regulation. New buildings would be made from large amounts of natural material acting as sequestration as well as providing low-carbon raw materials.

Transport would involve massive infrastructure change to public transport; there is high upfront costs but low maintenance. The car as a means of personal transport would diminish unless run on electricity.

The *landscape* would reflect changes in diet and fuel sources. Huge areas would be dedicated to carbon sequestration and meat consumption would shrink with the accompanying changes to the countryside.

Powering up of renewables is the key feature, with the bulk of electricity generation from offshore wind power, couple with a 50 per cent reduction in energy demand through efficiency measures.

For many of the sectors covered in ZCB the scenario is imaginable. *The Guardian* illustrated the ZCB report like this (Jowitt, 2010):

> Cars will be electric, and instead of owning them many drivers will borrow from car clubs or lease them. Airlines will no longer fly short distances and long haul trips will be a rare treat. Workers from more traditional heavy-energy industries like steel or cement will need to retrain to work in installing millions of buildings or back on the land, possibly involving big social upheavals. Dinner might be a roast, but poultry or pork because lamb or beef rearing would take up too much land and emit too many greenhouse gasses; while mango and bananas will be a luxury as food imports have halved. And the very landscape of Britain will look different too: instead of green and pleasant fields with grazing Friesian cattle there will be millions more acres of vegetables and grain to eat, and trees for biofuels or buildings.

Here we start to build up a picture of a zero-carbon society: its food, its industry, and its buildings. The vast arrays of offshore turbines. The fields of vegetables that have replaced barren green desserts of sheep and cows. Quiet roads and skies from the huge reduction in personal transport. But there are some aspects that are merely hinted at, 'possibly involving big social upheavals', says Jowit. As if this somehow sits comfortably alongside a pork roast dinner for all the family. We seem to have skipped a huge chasm of behaviour change and policy intervention.

The rest of this chapter focuses on the challenges of the unimaginable: the international agreements required to respond to the severity of climate change. The national economic and social interventions that may need to happen in the UK and the likely resistance that these measures will provoke.

International policy: The case for the UNFCCC

Once at the Centre for Alternative Technology, during a lecture to the Masters students we wrote down all the acronyms from the UNFCCC on flipcharts and stuck them round the walls. We filled sheet after sheet with unmemorable, dry names for countries, groups of countries, meetings and policy frameworks. Land Locked developing Countries (LLDC) got confused with Least Developed Countries (LDC) and barely anyone could even remember what the UNFCCC stood for, let alone the hundreds of supporting bodies. One umbrella group of countries is called just that – the Umbrella Group. Someone was convinced that the Major Economies Forum (MEF) was actually the Major Emitters Forum, ironically.

It is easy to see why the UNFCCC has a bad name. The endless meetings and arduous democratic process is still startlingly unequal in relation to the power of different countries. One noticeably significant difference is the entourage that comes with the heads of state. Most nations arrive with a small group of people, while financially wealthy countries like Britain fly in teams of researchers to help with the meeting processes. When optimisation of the processes relies heavily on the knowledge of up to date documents and issues, the extra people to read through and brief their leaders is essential. In the play *Greenland* (which addresses climate change issues) the character representing the G77 nations, knee deep in imaginary documents, says that delegates from the long industrialised countries like Britain and the US, take naps in their expensive hotel rooms right above the conference, while groups from the majority, less affluent world travel for miles to the conference with nowhere to recharge during the daily grind of endless negotiations. This way the people with the power retain the power.

Colin Challen, founder of the All Party Parliamentary Climate Change Group opens his book *Too little too late* with an entry that reads like a farcical play. In fact it's a real transcript from the Subsidiary Body for Implementation (SBI) at the Conference of the Parties (CoP) 13. As the players: China, Pakistan, Saudi Arabia, Bulgaria and Egypt painfully negotiate the tiniest of technicalities over a previous agenda point, the possibilities of misinterpretation are numerous. At one point Japan says (Challen, 2009):

'I left my memory stick in Nairobi, so have no memory...'
Chair '... some colleagues are in dissent, shaking their heads in the wrong direction ...
Pakistan: 'I'm shaking my head because I have a headache.'

In spite of his own criticisms Challen calls the UNFCCC an 'unsung hero'. This is mainly because of the feat of listing the carbon emissions of every countries on its website, but also because of its consensus-based decisions based on decentralised power. Due to its perceived lack of power, he calls the UNFCCC 'the weak king' surrounded by the 'strong barons' like the World Bank. Although riddled with historic and economic power imbalances; it's just that the long and thorough meetings for all parties prevent the full force of these imbalances from emerging. One can only imagine how different the international arena might be if there were no consensus process at all.

ZCB recommends a *global cap on carbon emissions* to keep the world within two degrees of warming that also addresses *international equity*. It is committed to climate science as the boundary for action, rather than economic or political realism, yet the policy solutions on offer are practical and decisive. Let's recap on the seemingly simple overview. A legally binding cap on carbon emission that keep the world within two degrees, requires a descent right now with global emissions peaking in 2016, as far as we know. The science may become clearer – the IPCC is notoriously conservative in its estimates – we may have to move faster.

Global equity: climate change is a social and political issue as much as an environmental one. In ZCB solving climate change becomes the opportunity to address intergenerational justice, historical emissions and outsourced consumption emissions. It also ensures that the people across the world affected by climate change receive adequate compensation from the long industrialised countries and British citizens are protected from price fluctuations, through economic instruments like taxation.

This is nothing new or radical. The original treaty articles of the UNFCCC that were reached by consensus at the Rio Earth Summit in 1992 state that (UNFCCC 1992):

The ultimate objective of this Convention and any related legal instruments that the Conference of the Parties may adopt is to achieve ... stabilisation of greenhouse gas concentrations at a level that would prevent dangerous anthropogenic interference with the climate system. Article 2

The Parties should protect the climate system for the benefit of present and future generations of humankind, on the basis of equity and in accordance with their common but differentiated responsibilities and respective capabilities. Accordingly, the developed country Parties should take the lead in combating climate change. Article 3 (1)

Challen says this is the bit we should not forget and 191 countries signed up to this treaty. ZCB offers three road maps for capping carbon through pricing mechanisms:

- Roadmap one: One price for all, which offers a reduced role for national and blocs of countries by agreeing a global carbon price within an international treaty, such as Cap and Share, Kyoto2 or a carbon tax. National governments, according to ZCB would still have to design policies to complement the agreement and 'balance the winners and losers form such a scheme.'
- Roadmap two: An international framework with national action initiatives. Carbon cuts for countries are agreed globally, but the mechanisms for achieving them are set nationally. The framework and cost allocations are via the global consensus process and the application at national and regional level allows policy flexibility.
- Roadmap three: Regional carbon pricing schemes. In this scenario, which ZCB describes as the most likely, countries would act regardless of the international framework, on their own, such as Britain with the Climate Change Act or as blocs, such as the EU Emissions Trading Scheme. Border adjustment taxes would prevent unfair competition, and national and regional carbon reduction targets could be set and enforced.

What seems very clear from all three scenarios is the high level of democratic state intervention, either at a global or national level.

Roadmap one is about pricing carbon at it true cost by the same amount across the world, supported by social payment transfers to distribute carbon 'wealth.' Roadmap two is still about setting enforceable frameworks while roadmap three sets carbon budgets with border taxes to correct issues like freeriding and outsourced consumption emissions.

These scenarios will only emerge, particularly in the long industrialised countries, if there is significant government intervention at global and maybe national levels in parallel with huge lifestyle changes from the electorate. When this is coupled with issues of justice and equity, both internationally and at home, the scale of intervention and education look very large.

At present it is hard to see how this leap can be made. Civil society and NGO pressure can achieve a large amount in terms of pressure on government policy, but still climate change is viewed by Neil Carter (2009: 119) as 'bad politics; a vote looser' and that 'There is a perception that the electorate are unwilling to accept environmental policies as they will impose significant costs or require lifestyle changes.' This view is probably replicated across the long industrialised countries and that perception is before we add climate change adaptation funding and other equity dilemmas into the mix.

An international climate agreement that makes sense to the world requires supportive policies at a national level. An international agreement that is not set by consensus as part of national and regional influence would merely be verging on a global dictatorship and probably impossible to achieve or enforce.

The international arena that enables an equitable, global carbon cap to be agreed is one where there is a high degree of democratic process. In the fictional scenario, envisaged by Alex Evans and David Steven (2009), an International Court for the Environment, based on the International Court of Human Rights, will increasingly enforce legislation on 'environmental crimes'. Countries from the majority world would have a greater influence over proceedings, and they may have grouped together for strength. We saw this in the emergence over the last decade of the African bloc and in the way that Bolivia vetoed the last international accord.

As the debate behind climate change becomes more urgent and its effects more keenly felt by the poorest in the world, an international agreement will emerge that has equity and justice at its centre. We may only feel the effects of climate refugee movement indirectly, as in the calls for reparations because of historical emissions, but it will be felt none the less. In spite of the tabloid hysteria when people move en masse it tends to be to countries nearby and sympathetic, but it will still be an ethical issue of how we as a richer nation will support those affected by our emissions.

The issue of consumption emissions will force lifestyle changes in the richer nations. Emerging now as an accounting issue, these carbon 'omissions' will need to be addressed. We may well find the business as usual approach impossible to rely on. As the years progress the

Transition countries like China and Pakistan will resent having to count such a huge quota of their emissions that are effectively embodied in cheap goods for the West.

An imaginary scenario of global proportions

It is 2040 and the over the last few decades the UNFCCC was strengthened by the emergence of new power from countries around the world. For various reasons, mainly due to energy shocks and financial blips, civil societies across the world and progressive governments created a replacement to the Kyoto protocol that had justice as its central premise. Some of the previous obstructors, like the US, having regulated their markets and invested hugely into renewable, evolved with the renewed, robust consensus process.

The International Court of the Environment initially began its life in response to a major obstruction by some of the Western countries back in the 2020's who refused to count their consumption emissions from abroad. The severe penalties for lack of accounting on such a serious issue – economic sanctions and military threats – were enough to set a pretty good example for the next decade or so.

Local Agenda 21 evolved into a more robust and democratic education programme. The youth organisations took the environmental issues very seriously and many of their recommendations on intergenerational justice became international policy. Their representatives met in-between the main negotiations, with delegates from schools and collages all across the world forming an environmental network. Schools elected their regional representatives who travelled to the meetings, often in training in careers in international politics.

Following this huge surge of civil society movements, international conferences addressed the need to update some of their procedures. One recommendation was that teams of researchers and civil servants from the wealthier countries were assigned to heads of state of poorer countries to help them in international negotiations, and thus to redress the global balance of power.

Britain: The zerocarbon policies

Within the international policy frameworks suggested by ZCB the role of national government is vital to encourage that collective global response. It assumes that national policies will allow an equitable, low-carbon transition, intergenerational, internationally and nationally.

It suggests that future generations of the world would be protected in environmental legislation as would people across the world that are more likely to feel the brunt of climate change. The transition in Britain to a low-carbon infrastructure would require domestic policy intervention to ensure protection.

As Roger Liddle and Simon Latham comment (2009: 63) 'Every effort must be made to financially support the vulnerable during the transition to a zerocarbon economy. ... The market is a powerful tool but needs to be regulated and controlled to ensure a socially just outcome.' They view the recent economic shocks as an opportunity to reinstate control and regulation of financial institutions, which in turn can be extended to other forms of regulation and state intervention, such as 'renewing the welfare state for low-carbon transition.' The 'crisis in market liberalism and the imperative of low-carbon transition should be viewed as a fortunate coincidence that will allow the state to recast itself thoroughly as both more active and developmental', they suggest.

The UK supposedly has the most progressive action and outcomes on climate change (as long as we are still not counting outsourced emissions) according to the government's *Low Carbon Transition Plan of 2010*. However we are still very far from accepting national policies that would really contribute to mitigating global warming. We are also, arguably about to head down a less regulatory route, with the Conservative led coalition government, in power at the time of writing. For how long, we cannot tell.

ZCB is clear that there are no technical barriers to an almost fully renewables based infrastructure and no significant scientific discord over climate change. So what is the barrier that is left, that produces so much political inaction? Back to Neil Carter's earlier mention of 'bad politics' and the perception of the UK voter (Carter, 2009). The barrier that seems so insurmountable: the behaviour of people that contributes to global warming.

Environmentalism, particularly climate change is described by some as a meta issue. Its solutions are claimed by many ideological camps that range from it being a symptom of economic market failure (the ultimate freeride) to the outcome of consumption-based capitalism. In some environmental discourses that problem is viewed as a symptom of civilisation itself, with no possible compromise between human societies and nature (e.g. by the groups Dark Mountain Project (Kingsnorth, 2010). Across different countries, policy direction is just as varied. A country such as Britain, may see climate change as an issue that requires a market solution (say through the ETS) right through to

Bolivia who appear to be leading the fight on behalf of countries in the majority world who feel that the economic markets of the long industrialised countries are the cause of the problem and need to be curbed.

At some point it feels as if we need to select a viewpoint on what the root cases of climate change are. If we do not do that, at least on a national level, a mismatch of methods from the differing ideological spheres will rub confrontationally alongside each other. Research evolves constantly and I do not imagine that the issues and solution I outline will remain the same for any length of time. But if we are taking the strap line of ZCB 'the science says we must, technology says we can, time to say we will' as the identification of the main societal barrier to change, the key issue now is of the constraints on actions, by citizens, policy makers and businesses, that actually stop climate change.

There is some evidence that suggests that the more democratic a country, the better its track record on environmental policies and in addressing climate change (Herd and Hervey, 2009). This may suggest that democracies based on a high level of learning and participation are better equipped to deal with environmental problems.

ZCB calls for a huge 'increase in public understanding' particularly of climate science. But even ZCB is tentative; the answers are more likely further questions to which there seem to be no simple solutions. Similarly to Randy Olson's book *Don't be Such a Scientist*, it suggests that the scientific jargon is too difficult to decipher and needs re-framing into language that the public understands, like 'carbon pollution is better than carbon dioxide emissions'. Olson (1997) calls it the type 1 and type 2 errors. Type 1 error is the lack of credibility if the academic language becomes to colloquial. Type 2 is the error of boredom as the scientific and academic language is lost on audiences. Science and academia is damned if it does and damned if it does not, he remarks.

What kind of policies would stop people from driving their cars? Over consuming products whose embedded carbon is counted elsewhere? What makes transferring social payments at home and abroad acceptable and welcome? Both the authors of ZCB and *Building a Low Carbon Future* suggest that the policy tail wags the dog, so to speak. Policy interventions and economic instruments like payment transfers to combat transitional fuel poverty help adjust attitudes to the underlying issues. In this case if the fuel poverty is caused by rising carbon pricing, it draws attention to both environmental issues and wealth distribution issues. Liddle and Latham (2009: 64) calls it 'subsidies that trigger behavioural change'.

To complement this a growing political and social movement may take hold and capture imagination: Tom Cromptom's research (2010), documented in 'Common Cause' is an exploration of the case for encouraging 'Bigger than self' values. In a similar vein Rosemary Randall, from the Cambridge Carbon Footprint charity, suggests dropping the term behaviour change altogether when it comes to environmental issues and seeing environmentally destructive behaviour as symptoms (Randall, 2011):

The behaviours that are targeted in climate change interventions – leaving appliances on standby, exceeding the speed limit, wasting food – are also symptoms. Behind them lie our disturbed relationship to the rest of the natural world, our fragile identities dependent on 'stuff', our anxious preoccupations with security and status. Impacting on each one are the decisions of governments and corporations, the predations of marketers, the values of the dominant culture and the opinions of our peers. And just as with psychological symptoms, if one behaviour is vanquished another pops up to take its place. In climate change work, this is what economists describe as the rebound effect. The money saved by insulating the loft and swapping out the light-bulbs results in the thermostat being turned up and the lights being left on for longer, or is blown on a flight to Madrid. The underlying problem remains.

A scenario: Britain – the just society

Its 2040 and many decades of cross party consensus, bolstered by the evolving climate science, meant that the carbon budgets set by the Committee on Climate Change in 2008 were strengthened. Instead of being target focussed (the UK seemed to keep falling short of its voluntary targets, as did many other countries) most of the EU countries collectively agreed on a 'race out of carbon'. The carbon pricing set by the new UNFCCC treaty was significant, but a huge civil society surge on the issue of environmental justice was also having an effect.

Once it became clear that countries were going to be required to count their outsourced emissions attitudes in the financially wealthy countries began to change. After significant backlashes in the tabloid media and protests at rising prices, a more egalitarian era was ushered in bit by bit. A large anti-taxation movement vied for political attention with the movement against poverty. Wealth redistribution became the issue of the 2020's and eventually higher taxation for investment into renewable infrastructure won the day. The example of some of our European neighbours led the way as well as the impact of higher carbon prices, priced at their true cost.

A threat of economic sanctions, when we tried to delay counting our outsourced consumption emissions was enough to shock us into more progressive policies on taxation and green investment. Coincidently, a highly charismatic head of state during the 2025–9 helped make the idea of social payment transfers very acceptable and for many years Britain was riding on a political wave of being a 'just society' that looked after the less well off and sought ways to redistribute wealth. In schools, an overhaul of the education system put Democracy, Ethics and Values firmly into the curriculum and many pupils studies Empathy as a subject. The inevitable tabloid backlash eventually died down after government perseverance.

Conclusion

The future is so dependant on the past and present; this home of our great grandchildren and beyond, does not belong to us. Yet we continually steal its resources as if it were a colony with no defences. We need to be its defence. We should be the guerrilla fighters and activists for the future. It is possible that the answer lies in mundane scenarios. Rather than the Earth First! style of arguably macho defence of the earth the solution may lie in the strengthening of democratic state intervention. The protection and strengthening of democracy may well be the policy silver bullet for addressing climate change.

ZCB argues for large amounts of state intervention to adjust carbon pricing either unilaterally or multilaterally, to protect citizens in Britain from fluctuating energy pricing, to stimulate the economy through public investment into renewable, and to transfer funds to countries struggling with the effects of climate change.

Tim Jackson (2009) in his book *Prosperity without Growth* lays out a 12-point plan to protect the Earth's resources and climate, nearly all involving a high degree of state intervention: investment in public sector jobs, higher taxation for higher earners, carbon budgets set by the Committee on Climate Change, tackling inequality globally. Similarly, Antony Giddens (2009) argues for return to long-term state planning.

These policy suggestions involve a huge reframing exercise. A paradigm shift from extrinsic values and self-interest policies. Although not necessarily a move to left wing ideologies – many extreme left wing regimes had a terrible record on environmental destruction (Cuba being a notable exception), I would argue that it requires a move to democratic state intervention and protection of the vulnerable: the vulnerable being the poorer people in our society and globally, the Earth's resources and the future.

References

Carter, N. (2009) 'Can the UK Reduce its Greenhouse Gas Emissions by 2050?', In: A. Giddens, R. Liddle and S. Latham (eds) *Building a Low-Carbon Future: The Politics of Climate Change*, London: Policy Network.

Challen, C. (2009) *Too Little Too Late: The Politics of Climate Change*, Uckfield: Sussex: Picnic Publishing.

Crompton, T. (2010) 'Common Cause: The Case for Working with our Cultural Values', Godalming, Surrey, WWF-UK. assets.wwf.org.uk/downloads/common_cause_report.pdf.

Daly, H. (1996) *Beyond Growth*, Boston, MA: Beacon Press.

DECC (2009) 'Low Carbon Transition Plan', London, Department of Energy and Climate Change'. www.decc.gov.uk/en/content/cms/what_we_do/lc_uk/lc_trans_plan/lc_trans_plan.aspx.

Evans, A. and D. Steven (2009) 'An Institutional Architecture for Change', London: Department for International Development.

Giddens, A. (2009) *The Politics of Climate Change*, Cambridge: Polity Press.

Herd, D. and A. Hervey (2009) 'Democracy, Climate Change and Global Governance', London: Policy Network.

Jackson, T. (2009) *Prosperity Without Growth: Economics for a Finite Planet*, London: Earthscan.

Jowit, J. (2010) 'Zero Carbon Vision Sees UK as Cleaner, Greener and Leaner within 20 Years', *The Guardian*, 16 June 2010.

Kingsnorth, P. (2010) 'The Dark Mountain Project', www.dark-mountain. net/2010.

Liddle, R. and S. Latham (2009) 'How Can the Response to Climate Change Be Socially Just', in A. Giddens, R. Liddle and S. Latham (eds) *Building a Low-Carbon Future: The Politics of Climate Change*, London: Policy Network (www.policy-network.net/publications/3136/Building-a-low-carbon-future-the-politics-of-climate-change).

Mckee, R. (1997) *Story: Substance, Structure, Style, and the Principles of Screenwriting*, London: Methuen.

Olson, R. (1997) *Don't be Such a Scientist*, Lyndhurst, NJ: Barnes and Noble.

OST (2006) 'Intelligent Infrastructures Futures: The scenarios – towards 2055', London, Office of Science and Technology. www.bis.gov.uk/assets/bispartners/foresight/docs/intelligent-infrastructure-systems/the-scenarios-2055.pdf.

Randall, R. (2011) 'Is it Time to Stop Talking about Behaviour Change', rorandall. org, 27 April 2011.

UNFCCC (1992), 'United Nations Framework Convention on Climate Change', United Nations, unfccc.int/resource/docs/convkp/conveng.pdf.

Wrigley, E. (2011) *Energy and the Industrial Revolution*, Cambridge: Cambridge University Press.

ZCB (2007) 'Zerocarbonbritain: An Alternative Energy Strategy', Machynlleth, Wales: Centre for Alternative Technology.

ZCB (2010) 'Zerocarbonbritain2030: A New Energy Strategy', Machynlleth, Wales: Centre for Alternative Technology. www.zerocarbonbritain.com.

11
Low-Carbon Society in Switzerland

Bastien Girod

Switzerland is unique compared to other European countries as there is hardly any heavy industry and electricity production is a nearly carbon-free, consisting of 60 per cent hydro and 40 per cent nuclear energy. As a consequence the greenhouse gas (GHG) emissions per capita are quite low, despite the high-income level. In 2007 the annual emissions (according to the Kyoto Protocol) amounted to 'only' 6.75 tons of CO_2-equivalent per capita (FOEN, 2009). However, this number does not include the whole carbon footprint caused by Switzerland. Many products manufactured or consumed in Switzerland are produced in other countries, while the exported goods and services do not use that much energy. As a consequence Switzerland has the highest per capita net imports of GHG emissions of any country (Hertwich and Peters, 2009). If the emissions embodied in imported goods and services are included, annual GHG estimates a range of 12–18 tons per capita (Jungbluth et al., 2007).

This chapter however focuses on the emissions according to Kyoto Protocol (territorial) measure, since these system boundaries corresponds to the system boundaries of national policy. Nevertheless embodied emissions must be kept in mind, especially to avoid measures that decrease emissions in Switzerland and increase the emissions elsewhere. To safely stabilise global warming below two degrees, GHG emissions have to be reduced to one ton per capita by 2050 (Meinshausen et al., 2009). Hence Switzerland has to reduce its GHG emissions at least by the factor of seven.

Figure 11.1 shows total emissions by sector from 1990 to 2007. Overall Swiss GHG emissions have decreased only 2.7 per cent. The Kyoto climate target has been met mainly by buying overseas carbon offsets. However, per capita emissions in the same time period have fallen by

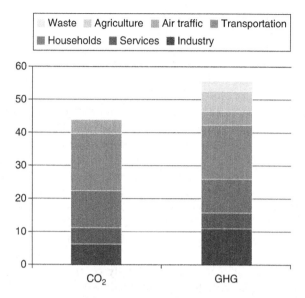

Figure 11.1 Swiss greenhouse gas and CO_2 emissions for different sectors

13 per cent, from 7.8 to 6.75 tons per capita since the Swiss population has increased (through immigration). The three most important GHG sectors are heating (in buildings), transportation and industrial processes. For non-energy related GHG, carbon dioxide emissions from cement production, and methane from agriculture are relevant. Overall a considerable decline in buildings has been offset by the increase in the transport.

Strategies to reduce CO_2

From Figure 11.1 the main strategies can be derived for reaching a low-carbon society in Switzerland. This chapter focuses on energy-related GHG emissions amounting to 77 per cent of total GHG emissions (including air transportation), knowing that in addition measures for agriculture (methane) and industry (CO_2 from cement) are also needed for reaching the two-degree target. Thus three main strategies can be identified to reach the factor seven reduction required for GHG emissions in Switzerland: zero-emission buildings, low-emissions transportation and green electricity supply. The (socio-) technical changes needed are first discussed, followed by the question of 'how to get there'.

Zero-emission living

Net GHG emissions from building can be reduced to zero through energy efficiency measures and renewable supply. The first step is to reduce building heat loss and this can be achieved by effective insulation, which includes avoiding 'warming bridges' (construction elements that conduct heat outside) and insulating windows. However with increasing insulation the heat loss through ventilation becomes more important and it becomes necessary to have controlled ventilation using a heat exchanger. Then the (reduced) energy use for the building can be supplied by solar energy technologies and geothermal heat from heat pumps, and houses can even become net energy producers. Figure 11.2 shows an example of an 'zero emission building'.

The photos show a large office building, using only a quarter of the energy of a comparable conventional building. A third of its electricity is provided by solar panels (Forum Chriesbach, 2009). Another pioneering building (not shown) is a multifamily complex, which produces more heat and electricity than is consumed by its inhabitants, allowing it to export energy. Furthermore many recycled construction materials and wood were used during construction of the building to reduce the embodied emissions (Jenni, 2009). Because of the long lifetime of buildings, the transformation of the existing building stock to zero-emission ones is crucial. The architect Beat Kämpfen (2009) has shown that this is often economically feasible, for he has managed to convert, at an affordable cost, a normal double family (duplex or semi-detached) house into one which supplies all of its heating and half of its electricity needs.

Low-emission mobility

Personal transportation accounts for the largest share of transport emissions. Thus in seeking to reduce emissions it is important to consider the purposes of travel. People do not travel for its own sake but for the activities and goals at their destination. Hence intelligent structure of settlements that allows daily activities within a short distance is crucial to reducing the total number of kilometres travelled. The structure of settlements influences the modal split and short distances between home, work and leisure activities allow people to travel by bike and on foot. For longer-distances public transport, if fuelled by carbon-free electricity, can reduce energy use and be zero-emission mobility.

The use of bicycles and public transport is, of course, heavily dependent on the quality of public infrastructure, and this can be shown

Figure 11.2 The Forum Chriesbach, a five-storey office and research building for EAWAG, the Swiss water research institute near Zurich (Duebendorf). The building for 220 persons uses no more external energy for heating and cooling than one to two single-family houses. The fresh air supply occurs through an earth tube collector that preheats the air in winter and cools it in summer
Source: Forum Chriesbach, 2009.

by comparing similar sized cities. Compared to Geneva, Zurich has an excellent public transport infrastructure, with comfortable trams providing access to all parts of the city. This results in Zurich having double the share of public transport compared to Geneva (ARE and BFS, 2000). However Zurich rates badly for bicycle use: bicycle lanes are rare and often missing at crossings. In contrast the nearby city of Winterthur (a third the size of Zurich) has its bicycle infrastructure rated top by cyclists (Pro Velo, 2005). Despite similar topographic, urban and weather conditions the bicycle share in Winterthur (12%) is nearly double that of Zurich (7%).

For car journeys that cannot be met by public transport (which will amount to a considerable share) zero-emissions vehicles are needed. This is technically most feasible with electrical cars. There is a continuous path from the combustion engine to the electric car, with the first step being the hybrid car that has already sold quite successfully. These cars have a combustion and electrical engine with a battery, which use the more efficient electrical motor for slow speed and acceleration as well as recovering braking energy. The next step is 'plug-in cars', which can be plugged into the electricity grid. At first they will have a combustion engine to extend their range, but with increasing storage capacity of batteries or a system to charge electrical cars very quickly (e.g. replacing the empty battery with a full one), no combustion engine will be needed. The zero-emission car will then become a reality. The electric car also has several other advantages, mainly no local air pollution, less noise and even using the batteries to stabilise the electrical grid from fluctuations in wind and solar energy.

Fuelling cars with bio-fuels is much less useful, since it has several risks and disadvantages. First, the benefits from the electric car (reduced air pollution and noise) cannot be realised. Second, the high demand for bio-fuels could drive food prices up if they are grown on land that is used for food production, and could also promote the clearance of tropical rain forests. There are some bio-fuels, which are less controversial, like those produced from organic waste. However, their potential is limited and hence they should only be used for transport modes, which cannot be replaced by electricity, like essential heavy freight transport (mostly trucks) and aircraft.

Air traffic probably presents the greatest challenge to reaching a low-carbon society, because of its projected steep growth in demand and technological solutions being limited. Only fossil fuels have the required energy density and can be stored safely enough (Schäfer et al., 2009). Even if sustainable bio-fuels are used non-CO_2 emissions like nitrogen oxides and water vapour remain that are estimated to heat up the atmosphere two to four times more than the CO_2 emissions alone (Kollmuss and Allison, 2009). The growth in air traffic is the reason why technically zero-emission travel will not be possible, so cutting growth in demand is crucial if we are to reach a low-carbon society by 2050.

Figure 11.3 illustrates the societal challenge of mobility. Technical improvements are often expressed in terms of CO_2 emission per person-kilometre. However, observation all over the world show that on average, people spend about one hour per day and ten per cent of their income for travel (Schäfer et al., 2009). As a consequence, with increasing income,

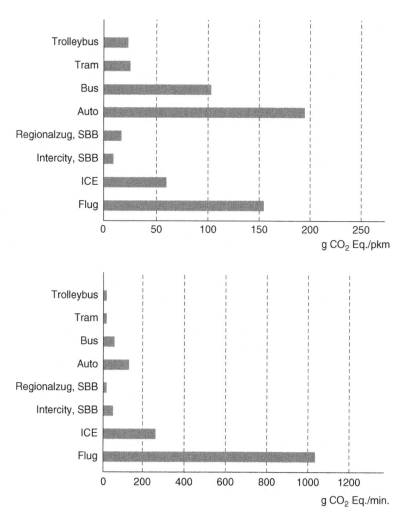

Figure 11.3 CO$_2$ emissions of different transportation modes
Source: Spielmann and Scholz, 2005.

there is a shift to more expensive and hence faster travel modes. Hence the emissions per time and not per person-kilometre should be used for comparing the carbon emissions of different transportation technologies. While person-kilometres will increase with income, the time used for travelling will remain more or less constant.

The measures for freight transport, which amounts to about a quarter of total emissions from transport, are similar to those for personal transport. However, the electrification of road freight transport is not possible because of the limited capacities of batteries. Hence, the shift to rail transport is more urgent than for personal transportation and the remaining trucks must be fuelled with sustainable bio-fuels.

To assess the climate impact of a single trip it makes sense to consider the environmental impact per person-kilometre (above). However, because the global average for travel is about 1.1 hours per day, the overall environmental impact is more meaningful expressed in terms of emissions per hour (below). [Translation of travel modes: trolleybus, bus, tram, car, regional train, intercity train, high-speed train, and airplane].

Green electricity

Today, Swiss electricity production is nearly CO_2-free (comprising 40% nuclear power and 60% hydropower). However, Swiss nuclear power plants have to be decommissioned at the end of their lifetime, which is within the next 20 years. Even if there is a tremendous push for energy efficiency, a reduction of electricity use by 40 per cent will not be possible. The policy of decarbonisation of the economy and transport sector will lead to additional electricity demand, due to the switching from fossil fuels to electricity. Hence, new electricity supply needs to be established in addition to the existing hydropower. Using nuclear energy to fill this gap is not acceptable to many people for two reasons. First, using nuclear is very contentious since climate risks are exchanged for nuclear risks and that of nuclear waste replaces the burden of a worse climate. Second, from a strategic point of view using nuclear energy for fuelling a low-carbon society could create large and strong resistance against the idea of a low-carbon society. The green movement – strongly in favour of a low-carbon society – might start to fracture. For a low-carbon society fossil-fuelled electricity production is not an option either. Hence 'green' (or renewable) electricity is needed.

Renewable energy is widely available and existing wind and solar technologies could provide the needed electricity supply. The limiting factors for its realisation are (i) market costs, (ii) grid capacity and (iii) supply security. The (initial) costs disadvantage can be overcome by financial incentives that would allow renewable to reduce costs through economies of scale and technological learning. Only then could they compete against fossil and nuclear energy that benefit from large installed capacity and absence of internalisation of environmental costs. Thus taxation of CO_2 emissions and nuclear risks would support such a transition towards green electricity supply.

In Switzerland wind energy potential is limited to about ten per cent of demand (Suisse Eole, 2010), whereas solar energy could provide 25 per cent of electricity if there were 20 square metres per person of photovoltaics (PV). However, the generation costs of PV are considerably higher than from wind energy from the North Sea or solar energy from the southern Europe or North Africa. Hence, an international grid, which could guarantee supply security, is very important for providing Switzerland with green electricity. The German Council of Environmental Advisors estimated that a hundred per cent renewable energy supply for Germany by 2050 would be feasible. However to reduce costs, it is important that countries with high hydropower capacities are connected to this international grid so they can balance out supply fluctuations from renewable with hydropower's pump storage (Sachverständigenrat für Umweltfragen, 2010). Hence, Switzerland can offer something to this European 'green' grid as its hydropower system can store and balance wind and solar energy.

How to get there

To reach a low-carbon society by 2050 fundamental changes in transportation, in living and in the electricity sector are needed. This section highlights the political actions or policies that are needed for such changes and the state of Swiss public opinion on such political actions. Liberal political parties, but sometimes also local movements, often appeal to consumers to change their behaviour in order to protect the environment. While this might be sufficient for some environmental actions, like recycling of waste, switching of the car engine at red lights etc., such behavioural measures are not enough to reach a low-carbon society. The citizen through political action, not the consumer through purchases, needs to be in charge for just appealing to consumers to change their behaviour, will only have a limited impact. This does not mean that consumer behaviour is not relevant. A considerable number of consumers have already reduced their carbon emissions by using efficient housing, little transportation, and spending money on quality instead of quantity (Girod and de Haan, 2009). However, further reductions are limited because of the increasing costs of additional measures and the unwillingness of the majority of households to voluntarily adopt such low-carbon behaviour.

Combining the higher costs of using green or low-carbon products for individual households with the positional utility of consumption, there is a prisoner-dilemma like situation. If everyone bought these green products, positional utility would not change. However, if only one person buys them the consumption level and thereby the positional utility decreases (cf. Jackson, 2009). In other words, if one buys a highly efficient car and house, and only rarely takes long-distance holidays, because of the higher costs the car is less large, the house is somewhat smaller and the holidays might be less exiting or exotic compared to the neighbour who does not buy more environmental-friendly products. Hence the utility from being better, or having more is lower. If all buy more environmentally friendly there is no such distortion. Similarly, entrepreneurs have only a limited impact as if they are the first to produce green products, that are more expensive than those of their competitors, they could lose market share. As for the consumer their actions will often remain in their 'comfort zone', which is however not enough to reach the low-carbon society. If entrepreneurs however put pressure on politicians to change the regulatory framework and internalise external environmental costs, green products can become more competitive.

Public support for political change to reach the low-carbon society is very relevant. Opinion polls show that Swiss citizens are very concerned about environmental issues. An example of their views was given by the results of a 2008 referendum in Zurich, the largest city of Switzerland. They decided by a large majority to incorporate into the municipal code the target of reducing annual GHG emissions to one ton per capita by 2050. This decision was linked in the same vote with the decision to phase out nuclear energy by 2050, and in addition reduce energy consumption to 2000 Watt per capita from the current 6000 Watt (Stadt Zürich, 2010). Also 130 of the 200 members of the Swiss parliament campaigned in the 2007 election to support a 30 per cent reduction of GHG emissions by 2020 compared to 1990. And in the first vote in the parliament on the post-Kyoto emissions reduction target there was a (very narrow) majority to support a reduction of GHG emissions by 2020 of at least 20 per cent within Switzerland compared to 1990, and up to 40 per cent if there is an international, legally binding agreement with other countries pledging also for more than a 20 per cent reduction.

While there is strong support for the idea of a low-carbon society, support declines for effective measures. A recent survey, sponsored by BP, shows the typical public attitudes towards mitigation measures for GHG emissions, which are in line with other surveys in other countries. Nearly 70 per cent opposed a carbon tax on gasoline (petrol), while subsidies for more 'green' cars (hybrid- or electric cars) were most popular with 60 per cent support (Marketagent.com, 2010).

Most probably, support for a CO_2-tax on gasoline would increase if it was proven to be the only way to reduce emissions, but it would be at the cost of reduced public support for the idea of a low-carbon society. So the social and political challenge is to find measures that are not only effective in reducing CO_2 emissions but also popular, and the rest of this chapter focuses on such measures in Switzerland.

Effective and popular policy measures

If people support the idea of a low-carbon society but are not willing to pay a high price for it, how can it be achieved at a minimal cost? First we must ask what are the costs? Looking only at monetary costs is attractive from a scientific and economic point of view. However, this relies on the assumption that the best solution is the one with the lowest monetary costs, and therewith allowing the highest consumption level. But for many people, other criteria like fairness or effectiveness are important. Moreover, the perceived and the real costs need not be the same; as Kahneman and Tversky (2000) showed in various experiments, people give higher weight to losses than to gains. All such criteria are implicitly considered in a democratic process. Thus the evolution of the Swiss climate policies may indicate to other countries which low-carbon measures are not only effective but also popular.

The story of the Swiss CO_2 tax

The story of the Swiss CO_2 law is a lesson about a scientific proposition meeting political realities. Or about objective monetary costs meeting perceived costs. From a scientific and economic perspective taxing CO_2 emissions was estimated to be the most cost-effective way to reduce carbon emissions. Based on these findings the CO_2 law of Switzerland in 2000 had as its main measure a fiscally neutral CO_2 tax. The tax was designed to raise carbon prices and redistribute the returns equally to businesses and population (on per capita basis). It envisaged a reduction of emissions of ten per cent (15% for emissions from heating, and 8% for transportation). Businesses could exempt themselves from the tax if they reached the CO_2 reduction goal through other measures. However, the law was still open to change and as a result the CO_2 tax was not implemented as originally intended.

The tax on gasoline (petrol) was never raised. Instead the proposal from the Swiss oil industry was accepted, which raised a very small tax that bought carbon offsets to compensate emissions abroad. For buildings however a carbon tax was introduced successfully. It was first introduced in January 2008 amounting to 12 Swiss francs (18 Euros) per ton of CO_2. In 2010 the tax was raised to 36 Swiss francs per tons and the actual post-Kyoto CO_2 law allows the government to increase the tax up to 60 francs per ton of CO_2. The broad acceptance of this carbon tax however, only began in 2009, when it was decided by the parliament that part of the tax could be used to support a programme for retrofitting buildings, which was supported by many business groups who would benefit from it. From an economic perspective using a fiscally neutral tax would allow the market to decide on retrofitting and hence is more effective than subsidising a retrofitting programme. However, the retrofitting programme leads to visible gains at low-tax rates. Without the retrofitting programme the tax would probably not have been raised from 36 to 60 francs. Thanks to the retrofitting programme this increase meant also additional visible gains, since the financial support of retrofitting could be enlarged.

Other measures were less controversial and effective. The increase in the efficiency standard for buildings reduced annual heating requirements to 4.8 litres heating oil equivalent (48 kWh) per square metre of living space. In 1975 buildings in Switzerland used an annual 22 litres heating oil equivalent per square metre. Hence the continuously increasing efficiency standards for buildings reduced energy use by the factor of 4.5 in 35 years. Also minimum efficiency standards for appliances were introduced. There was only little opposition and public controversy on these standards, even if they did, in some cases, lead to an increase in costs for buildings or appliances.

Countering car 'irrationality'

Standards for cars limiting non-CO_2 emissions, like NOx, particulate matter and other air pollutants, already exist. Standards for CO_2 emissions were mandated in a referendum that I supported. These standards led to a heated national debate, since a strict CO_2 per kilometre measure would have resulted in the banning of some SUVs and sports cars. This debate showed that measures to reduce carbon emission from cars probably incite stronger feelings than in any other field of climate mitigation, since people have an emotional attachment to their cars. Thus in determining national transport policy the challenge is to deal with these emotional feelings and to investigate measures leading to less resistance. For the Swiss post-Kyoto climate policy a new approach was taken, as a counter proposal to the popular referendum described above. Similar to the European Union policy, the idea is to reduce average CO_2 emissions of new cars by giving targets for reductions in average emissions of sold cars to the automobile industry (in Switzerland for automobile importers). If the emissions are above target fines have to be paid. The European Union decided to reduce average emissions to 130 grams per kilometre (g/km) by 2015. At first the EU envisaged a more ambitious goal, but due to successful lobbying from automobile industry the target is now less ambitious. The National Chamber decided to reduce the emissions also to 130 g/km by 2015, which is slightly more ambitious than EU because Switzerland starts from a higher emission level (in 2009 the average emissions were 175 g/km in Switzerland and 153.5 gCO_2/km g/km in the EU). Thus at least a measure has now been found, which allows the effective reduction of the carbon emission of cars and causes less hostility.

The success story of the feed-in tariff

Finally, the third key policy for low-carbon emissions is 'green' electricity. A very successful measure is the feed-in tariff for renewable electricity supply, which guarantees producers a fixed price for the electricity produced over a period of 15–20 years. The future cost of for new renewables can be reduced through exploiting economies of scale, with these costs met by a small increase in the average electricity price. Thus non-renewable energy is in effect 'taxed' although the price increases for non-renewable sources is low. In Germany it was 1.1 cents per kilowatt-hour (kWh) in 2007, in Switzerland the maximum is set below one cent per kWh (Bundesministerium für Umwelt, Naturschutz und Reaktorsicherheit, 2007).

The feed-in tariff was introduced in Germany in 1990, resulting in a boom in wind energy. Latter the increase in the minimum price paid for electricity produced from photovoltaic (PV) solar panels made Germany the world's largest solar electricity market. Meanwhile this measure has been copied worldwide by over 40 countries, and hence has probably become the world's most successful carbon reduction measure (Bundesministerium für Umwelt, Naturschutz und Reaktorsicherheit, 2007). Its popularity may be due to the public visibility of the gains (subsidies to homeowners with an expanding PV industry and growing employment in renewable energy production), and the costs being hidden (in the form of a small increase in average bills).

In 2008 Switzerland introduced the feed-in tariff for electricity produced from wind, solar, agricultural biomass and geothermal energy, with different fixed prices for each technology. Parliament decided to cap the total solar investments eligible for the feed-in tariff, since it is the most expensive. The limit for investments in solar energy was reached on the first day of the introduction of the tariff, which shows how the potential of solar electricity and the motivation of enterprises to install solar panels on their roof were greatly underestimated. Since then the limits have been increased. Even some conservative and pro-business parties, who normally reject climate mitigation measures, support this tariff since their supporters – farmers, house owners and businesses – can all gain from it. The success of this tariff, to me, illustrates how the creativity and the courage of a single country (in this case Germany) in pioneering a policy, is crucial to the world finding effective and popular measures towards a low-carbon society.

Effectiveness and rebound effects

Rebound effects must be considered when assessing the effectiveness of energy policies, as the general promotion of energy efficiency might lead to economy-wide rebound effects (Sorrell, 2007). However, measuring the changes in the GHG emissions monitors the effectiveness of policies. Figure 11.1 shows the effectiveness of the climate and energy efficiency policy. For buildings the CO_2 tax and efficiency standards were implemented, but for transport no measures were applied. Accordingly emissions of the buildings decreased, while those of the transportation increased. Rebound effects in climate policies exists when a broader environmental perspective is considered, and a short overview is in the following points:

- Carbon leakage: This is probably the greatest challenge for climate policy. If industrialised countries cut their emissions first this could simply reinforce a shift in their economies towards services rather than industry. However, their consumption of goods would still increase, but production now happens in other countries and is therefore not accounted in the national climate policy targets (e.g. Peters and Hertwich, 2008).
- Bio-fuels: These can also be a carbon leakage problem. Today's bio-fuels lead to GHG emissions from land use change but these emissions are not accounted for in countries importing such fuels. In addition bio-fuels could lead to many other negative social and environmental effects.
- Nuclear: Many electricity suppliers in Switzerland promote nuclear energy as climate friendly energy. A global revival of this technology could bring severe risks for human well-being.

- Negative side effects: Renewable energy can have negative side effects such as noise (from wind energy), disfigurement of landscape (from solar and wind), biodiversity loss (from hydro power, wind and bio-fuel plantations) and even earthquakes (from deep geothermal energy). However, these side effects can be strongly mitigated. The challenge here is that the lowest cost renewables are often not those with the lowest negative side effects.

All these possible rebound effects (or unintended negative side effects) of climate policies can– at least theoretically – be handled by good policy and fiscal measures. They however show how our growing economic system needs increasingly complex policies, to avoid negative feedbacks to our environment and society.

An iterative process towards the low-carbon society

To reach the low-carbon society in Switzerland GHG emissions have to be reduces by a factor of seven by 2050. This is quite a challenge even if technically feasible. The three major steps required are: zero-emission buildings, low-emission transportation and renewable electricity supply. Low-carbon buildings, cars and electricity already exist today and could be applied across the whole economy. However to do so, public support is needed to reinforce the political efforts for a low-carbon society. Saint Exupéry once said (de Saint-Exupéry, 1976): 'If you want to build a ship, don't drum up the men to gather wood, divide the work and give orders. Instead, teach them to yearn for the vast and endless sea.'

This citation is perfectly true for the low-carbon society and sustainable development in general. Most important, is to get broad support for the goal of reaching this society. In Switzerland public support for this goal is quite broad; in Zurich, the largest Swiss city, more than 75 per cent supported a referendum with such a goal. However, it is also important to realise the magnitude of the challenge ahead. Like one cannot walk on water but needs a good ship to meet the challenges of the sea, effective measures need to be at hand. A carbon tax would be a simple and very effective measure to reach such a society, and is probably the most recommended measure from the scientific and economic community. However it is an unpopular measure, just as are many other (cost) effective measures. Hence focusing only on such measures will reduce public support for the goal of a low-carbon society.

Instead climate policy should be seen as a continuous search for (technically) effective and popular measures. In Switzerland this search lead to different forms of financial incentives, subsidies and efficiency standards. I conclude from this process that one should not cling to a particular measure but be open to different approaches. Therefore the monitoring of the effect of different measures is most important. This allows an iterative learning procedure towards effective and popular measures. For the European Union and other countries willing to head towards a low-carbon society, there should be flexibility in the implementation of climate goals, and a diversity of approaches should be encouraged.

References

ARE and BFS (2000) 'Mikrozensus zum Verkehrsverhalten', Bern und Neuenburg: BBL.

Bundesministerium für Umwelt, Naturschutz und Reaktorsicherheit (2007) 'EEG-Erfahrungsbericht 2007'.

FOEN (2009) 'Switzerland able to meet its International Commitments'. http://www.bafu.admin.ch/dokumentation/medieninformation/00962/index. html?lang=en&msg-id=30656 [Accessed 24 July 2010].

Forum Chriesbach (2009) 'Kennzahlen', www.forumchriesbach.eawag.ch/ kennzahlen.htm.

Girod, B. and P. de Haan (2009) 'GHG Reduction Potential of Changes in Consumption Patterns and Higher Quality Levels: Evidence from Swiss Household Consumption Survey', *Energy Policy*, 37(12): 5650–61.

Hertwich, E. and G. Peters (2009) 'Carbon Footprint of Nations: A Global, Trade-Linked Analysis', *Environmental Science and Technology*, 43(16): 6414–20.

Jackson, T. (2009) *Prosperity Without Growth: Economics for a Finite Planet*, London: Earthscan.

Jenni, J. (2009) 'Kraftwerk Bennau – Projekt mit Zukunft'. www.jenni.ch/pdf/ Kraftwerk_Bennau2.pdf.

Jungbluth, N., Steiner, R. and Frischknecht, R. 2007. Graue Treibhausemissionen der Schweiz 1990–2004, Bern: Bundesamt für Umwelt.

Kämpfen, B. (2009) 'Kämpfen für Architektur – Energetische Erneuerung Doppeleinfamilienhaus Zürich, 2008/2009', http://www.kaempfen.com/index. php?option=com_content&task=view&id=144&Itemid=228 [Accessed 19 August 2010].

Kollmuss, A. and M. Allison (2009) 'Carbon Offsetting and Air Travel', Part 2: Non-CO_2 Emissions Calculations, Somerville, MA, Stockholm Environment Institute.

Marketagent.com (2010) 'Repräsentative Umfrage bei Schweizer Automobilisten zeigt', http://www.bp.com/genericarticle.do?categoryId=4004181&contentId= 7059391 [Accessed 25 July 2010].

Kahneman, D. and A. Tversky (2000) *Choices, Values, and Frames*, Cambridge: Cambridge University Press.

Meinshausen, M., N. Meinshausen, W. Hare, W. S. Raper, K. Frieler, R. Knutti, D. Frame and M. Allen (2009) 'Greenhouse-Gas Emission Targets for Limiting Global Warming to 2°C', *Nature*, 458(7242): 1158–62.

Peters, G. and E. Hertwich (2008) 'CO$_2$ Embodied in International Trade with Implications for Global Climate Policy', *Environmental Science and Technology*, 42(5): 1401–7.

Pro Velo (2005) 'Veloklimatest: Winterthur Top', Zürich Flop. www.provelozuerich. ch [Accessed 19 August 2010].

Sachverständigenrat für Umweltfragen (2010) '100% erneuerbare Stromversorgung bis 2050: klimaverträglich, sicher, bezahlbar'.

De Saint-Exupéry, A. (1976) *Die Stadt in der Wüste*. Düsseldorf: Karl Rauch Verlag.

Schäfer, A., J. Heywood, H. Jacoby and I. Waitz (2009) *Transportation in a Climate-Constrained World*, Cambridge, Mass.: The MIT Press.

Sorrell, S. (2007) 'The Rebound Effect – an Assessment of the Evidence for Economy-Wide Energy Savings from Improved Energy Efficiency', London: UKERC, www.ukerc.ac.uk/Downloads/PDF/07/0710ReboundEffect/ 0710ReboundEffectReport.pdf.

Spielmann, M. and R. Scholz (2005) 'Life Cycle Inventories of Transport Services', *International Journal of Life Cycle Assessment*, 10(1): 85–94.

Stadt Zürich (2010) 'Nachhaltige Stadt Zürich – auf dem Weg zur 2000-Watt-Gesellschaft', http://://stadt-zuerich.ch/portal/de/index/politik_u_recht/ politik_der_stadt_zuerich/legislaturschwerpunkte/2000-watt-gesellschaft.html [Accessed 24 July 2010].

Suisse Eole (2010) 'Potential Windenergie Schweiz'.

Part III
Stories

12

Little Greenham

James Goodman

To the unsuspecting visitor, Little Greenham might not look like much. It's not the prettiest village in the shire, with a major road running right through it. The village has built a large housing estate and a wind farm in the past decade and there are plans afoot for even more new homes. But this village of 2500 people – up from 1900 just ten years ago – has beaten national competition to come first in the Calor Village of the Year Award, 2016 and wins €50,000.

Little Greenham – Fact File

Population	2,500
Annual CO_2/head	zero
Cars per household	0.7
Local food consumed	52%

Sense of identity

Competition judges are looking for more than thatched cottages and tidy rose beds. 'We're interested in communities that have worked together to overcome the challenges of rural life', says judge Jeremy Taggart. 'Villages where people feel they belong, where there is a sense of direction for the village, and an ambition for where it could be in ten or twenty years time.' Little Greenham has that in spades. The village has been at the centre of a whirlwind of change in the past decade, and it's been vital to cultivate a sense of identity. 'People are proud to come from Little Greenham,' says villager Darren Andrew, 14, sporting a 'Little Greenhammer' t-shirt left over from the last village festival. Mavis Butcher, 81, a resident at the recently renovated elderly care home, says this is down to the hard work of the parish council, which has taken full advantage of the new powers made available to parishes back in the two-thousands. So, how has Little Greenham done it? One reason is that the village is so forward looking. 'We've got a vision, written down, of where we want to be in twenty years' time. It took ages to put together and practically everyone in the entire place was involved in one way or another', says resident Thom Lazzier. The vision is featured on the interactive touch-screen notice board next to the bus stop, as well as online. The competition encourages forward-looking vision, this is one of the set criteria for judging villages. Other criteria include participation, business opportunity and sustainability. Little Greenham was one of 3000 villages that entered the competition this year – three times the number in 2006 – and won the county prize before being put forward to the national competition.

The Toad Show

In their report on Little Greenham, the judges drew particular attention to the Toad Show, a festival that originated as a sausage making competition ('toad' is a local name for sausages). The festival died out in the mid-twentieth century, but the parish council revived it in 2012, and it is now thriving. 'Twenty-thousand people came to last year's Toad Show over three days', says festival organiser Amy Birtwhistle. 'Of course it's about much more than sausages. There's a funfair, music and a great deal of lovely, locally produced food.' According to Mrs. Birtwhistle, the festival has put Little Greenham on the map and attracts visitors from across the region. It has also helped to revive some of Little Greenham's traditional food production. A five-acre field has been set aside by a local arable farmer to grow vegetables for the show and for the village. Villagers pay an annual subscription that gives them an unlimited amount of vegetables for their own consumption. Thirty families are now self-sufficient in potatoes, carrots, cauliflowers and onions. Because of the Toad Show and the vegetable field, the village is well on the way to attaining the Government's goal for British villages to be 60 per cent self sufficient in food. But a few years ago the picture was not so rosy. 'Toads go back a long way at Little Greenham. The only pub in the village is 'The Toad and Turbine' and the old manor house, now an elderly care home, is still sometimes known as Toad Hall', explains local butcher Liz Burley, 49. But her butchers shop closed in 2007, and she remained unemployed until the Toad Show was revived. 'In that year, I signed a deal with the main pig farmer to be the exclusive sausage maker using the Little Greenham pig breed', says Burley. The business is now thriving: sausages are sold at the farmers' market that tours local villages and towns, as well as over the internet, and are delivered to people around the county as part of a local organic box scheme.

Local opportunities

Liz Burley's business has been helped along the way by an innovative project to support small businesses, called the Little Greenham Share Scheme. Shares of a new company are sold and shareholders receive dividends not in money but, in the case of Burley's Butchers, in sausages. The scheme has given villagers – and non-villagers too – a stake in the success of a local venture. It also encourages villagers to buy more sausages and to promote them among their friends and family. For Liz Burley and John Donald, the pig farmer she works with, the scheme has made all the difference. Money from selling shares was supplemented with a grant from the parish council. John Donald, 42, says, 'a few years ago, I was in serious debt and on the verge of either losing my farm, or taking European funds to grow biofuels. I wanted to stay growing food on my farm. Because of the festival, the share scheme and Burley's sausages, I could'. The same scheme has been used by others. For example, Jenny Martin has set up an energy microgeneration consultancy advising on how homes and small businesses can reduce their energy use. Shareholders receive discounted advice on their energy use and have first refusal on buying cheap, locally produced energy.

Little Greenham – what's new?

- Village vegetable field
- Little Greenham Share Scheme – support for new business
- Little Greenham wind farm – owned and run by the village
- Electric car pool
- Little Greenham II – twinned with Harapur II in India
- 300 eco-homes, 40% of them affordable
- Smart card for village discounts and data tracking
- Greenham District Railway Centre – railway station and telecentre
- Solar-powered village hall

Changing faces

Little Greenham's proximity to Big City has presented the village with some of its greatest challenges. The regional spatial plan earmarked Little Greenham as a 'growth village' and since then building work has been virtually non-stop. One large housing estate has gone up, and another is on the way. The parish council approved the first development of 300 houses, but using new powers insisted that 40 per cent of the new homes were affordable and made available to local residents first. The council also ensured that the houses were completed to the highest possible environmental standards, so that average CO_2 emissions per head in the village did not increase. The second, much larger development was opposed by the council, but driven through anyway. Councillor Pam Isleworth, 39, says, 'we just felt it was too much, too quickly. We needed time to adjust to the changes.' But once the new estate looked inevitable, the community decided to welcome the new homes as an opportunity. Councillor Isleworth again, 'we just knuckled down and made the best of it. Little Greenham won't be so little anymore, but we have no regrets'. The village runs regular 'new faces' evenings at the village hall, and every new arrival receives a 'Welcome to Little Greenham' pack 'not just stuffed through the letter box but delivered by hand by a neighbour'. The welcome pack includes a Little Greenham smart card, used for discounts with buses and taxis, and to register CO_2 emissions and the amount of food being bought from local sources. Real efforts are made to include everyone in decisions that affect the village, with regular opinion polls on local matters. And, as competition judge Jeremy Taggart says, 'there's a combined crèche and oldies club at the village hall to make sure that everyone can participate in village social life. It's incredibly important to village life that people are included, no matter their age or background'.

Going places

Many of the new Little Greenhammers work in Big City, which normally would be a problem for the village CO_2 emissions. But these are kept to a minimum: there's an electric car pool, managed online, and this has cut car ownership from an average of 1.4 per household in 2006 to 0.7 per household today – much lower than the average for rural areas. For non-drivers, there is a small electric taxi service run by volunteers, taking pensioners and the village youth to Nearby Town and back, free of charge. And then there's always the train station, eight miles away by bus. In another striking victory for local activism, the station was reopened after literally decades as a little-used warehouse followed by dereliction. A consortium of local villages previously served by the station lobbied the Shire Trains Company, eventually signing a deal in which the consortium guaranteed a certain number of passengers. Put together with match funding from the Regional Development Agency, this is enough for the four-times-a-day service to break even. According to Jerzy Dobavic, 38, manager of the Greenham District Railway Centre, the deal would never have been signed without renewed backing for rail from the Government. But it was the enthusiasm of the village consortium that had most impact, he says: 'We had to guarantee that people would use it, and to be honest we weren't confident we could do that. We had to make sure, and that's one of the reasons why we combined the railway station with a telecentre, drop-in office and meeting place. Now most weekdays we're full. A handful of small businesses are based here and many others come just to use our video conferencing kit – they love it.' This approach to local transport has been copied by other villages in the area and further afield, whether there's a train station nearby or not. Telecentres work at bus stations just as well. Even when at work in Big City or elsewhere, Little Greenhammers are never more than the click of a mouse away from their village. Little Greenham II is the online, virtual replica of Little Greenham, where people can meet, catch up, shop, do business or just find out what the village is up to.

Fit for the future

Not only has the village changed, but the world around the village is changing too. Rainfall in the Little Greenham area has been below average every year for the past eight years. In 2010, just 40 per cent of expected rain fell. At the same time, torrential downpours are more common, meaning that less water is absorbed into the ground. The village has had to adapt to near-drought conditions, and to keep gardens growing every single house has a water butt attached to the guttering to collect rainwater for reuse. There is also a new village pond, designed to be as resistant as possible to drought. 'Our pond does everything the experts say it should to conserve water. We're very proud. It's deep, is surrounded by bushes and long grass and it's right in the shade of trees', says resident Arnie Tibble. The pond attracts a variety of wildlife, especially during dry summers. 'Last year, we even spotted a pair of wild beavers – hundred of miles from where beavers were first reintroduced!' Little Greenham has responded vigorously to the challenge of climate change. In 2011, it won an award for going carbon neutral. This was achieved through a combination of energy efficiency in homes, use of renewable electricity and finally through the purchase of carbon offsets. But today there is no need for offsets: the village is not only self-sufficient in energy, but sells electricity back onto the national grid. Councillor Isleworth says 'I firmly believe that progressive villages should be showing the way to the nation on adapting to and mitigating climate change. We have the opportunity to be completely self-sufficient in energy, using natural resources such as waste from agriculture. We wanted to be an example and that's why we spent so much on the wind farm.' There was oppostion to the wind farm, but now it's more accepted. Arnie Tibble: 'Lots of us were sceptical to say the least about the wind farm. A lot of money for little benefit and a total eyesore. I still don't think it's pretty but I agree with most Little Greenhammers now – it makes us feel safe. We own it, we control it and it means we don't worry so much about getting the energy we need.' Every year the villagers conduct a survey, helped with information from the village smart cards and the Little Greenham II website, to estimate CO_2 emissions per person. The survey includes all local businesses and takes into account travel and shopping. The village hall was renovated in 2012 using national lottery funding. The latest, most efficient solar panels, using nanotechnology, were fitted on the roof and on the south wall, and now the village hall is able to sell energy back onto the national grid. The revenue from this innovative venture, along with money from the wind farm, goes directly into parish coffers, and funds other local environmental activities. One such project was the planting of a village wood, on a scrap of semi-derelict land once occupied by an abattoir. The woodland, two acres in area, is free to access for all, including the pigs of John Donald's farm, which enjoy the yearly crop of acorns. Little Greenham is a case study in how villages can adapt to a changing world and still be vibrant, prosperous communities. But this year's Calor Village of the Year is far from precious about its success. Visits from villages across the county and further afield are common: Little Greenham is spreading the message. And to cap it all, Councillor Isleworth announced today that Little Greenham II, the virtual counterpart to the real village, is now twinned with a similar virtual village, Harapur (meaning Green Village in Hindi) in rural Northern India. As Thom Lazzier says, cradling a pint of beer outside the Toad and Turbine, 'it's not paradise – nowhere is. No force on earth could separate my cousin from his vintage Ford Mondeo! But it is a great place to live. I wasn't sure at first about entering the Calor Village of the Year award, whether the effort would be worth it, but it's done us the world of good, it really has. It's helped bring us together, and realise what we've got.' Which, all in all, is quite a lot.

Note

This is a reprint of the publication *Calor Village of the Year Awards: Towards 2016* by Forum for the Future, see www.forumforthefuture.org/library/calor-village-of-the-year-awards.

13
The Housing Ladder

Roger Levett

Suppose the ever increasing frequency of catastrophes – floods, droughts, storms, landslips, crop failures, transport disruptions and so on – finally persuaded public opinion that climate change warranted serious action, and that the UK really did set itself on a path to low-carbon living and not just talking about it or expecting (or bribing!) others to do it for us. What then would housing, and people's attitudes and behaviour about it, be like? Let us peek into the home of a typical young English couple of 2050? Miriam Olowayo-Bull and Chang McPherson-Mohammed. It's a modest brick house in the middle of a terrace of similar ones in an inner city. At a first glance, they look much as they – and hundreds of thousands of similar ones – had done around 2000. At a second glance, surprisingly, they look even more the way they did when newly built in the late nineteenth century. Let's take a closer look and see why.

The terraced house

The clumsy, ill-proportioned replacement windows have all gone, and the original elegantly minimal sashes are back. Except these are not the originals, but modern ones with low emissivity triple glazing, fitting snug and draught-free on teflon runners, opened and closed by electric motors controlled by the house's building management system. Each window has a tiny symbol discreetly etched into one corner: sometimes a circle with a three bladed propeller in it; sometimes a little v above a w, sometimes a pair of chevrons. Because guess which industry leapt in to exploit the lucrative new market created by stringent regulations requiring energy efficiency retrofit of all housing without compromising their appearance when the traditional replacement window companies grumbled that it could not be done?

Also gone is the hotchpotch of replacement roofs. They are now identical and in perfect repair, just like when they were built. But those are not slates, they are photovoltaic panels, installed by the community energy agency. Participation was voluntary, but no households declined, because the new roof was free, and came with a Greenhouse Tax Credit that halved an average household's tax bill.

The coal plates are back in the path outside every front door. But they are not for coal, but to give access to the eco-utilities connections that now occupy the coal hole. Every house now has drinking water in (a micro-bore pipe), washing-grade water (a bigger pipe because it's a tenth the price, and with a two-way meter, because the same system is used to garner in rainwater from the house roof for central storage in rainy periods, and to supply it back in dry), biogas (from the local digester: another very narrow pipe because it's a premium fuel used only for cooking on the hob where its flexibility is really needed), hot water (from the local combined heat and power plant called Old McDonald's after the traditional drive-thru fast food place it replaced), grey water out (to the reedbeds), sewerage out (to the digester), and electricity (in or out depending on whether the roof is generating less or more than the household is using).

Outside, the public road does not look anything like how it did in either Victoria's or Elizabeth's reign. The road itself has been reduced to a single-track gravel path winding past a succession of narrow lawns, play areas, ponds and the occasional vegetable patch reclaimed from the previous parking. Traffic is back to a Victorian scatter of bicycles and small service vehicles: the late twentieth-century car infestation turned out to be a transient anomaly. However the service vehicles are now mostly battery-electric food delivery/recycling collection carts operated on foot, with the occasional electric taxi or minibus. Now look behind the houses: is that a lemon tree holding up one corner of the graceful canvas awning that shades the half of the back garden nearest the house?

Now let's go indoors. Or at least, into the inner end of the living space that merges with the garden, the bit that's inside the storm shutters when they are closed. Under the breakfast things, the LCD surface of Miriam and Chang's new Ikea digital kitchen table is strewn with estate agents and developers' e-brochures. Their first baby is on the way, and they are thinking of moving house.

Slumburbia

Chang flicks the surface of the table with the DigiThimble on his index finger, sending a copy of the brochure he's reading spinning across to

Miriam's side of the table, where it shunts the one she was reading out of sight under her papaya plate. She looks up reproachfully.

'Don't be prejudiced, love, just look at this one? It's the best of the lot. Please?' Chang implores. Reluctantly she taps one of the animated thumbnail pictures. It instantly enlarges to a view of a large detached house. Using the NaviPad that has appeared in the corner of the picture, Miriam pans slowly round, revealing a succession of similar large houses in a variety of period dress: a classical porch stuck on this one, leaded windows on that one, a bit of half timbering on the next. Her lip curls at the jumbled pastiche and the arbitrary, meaningless siting. Chang has clicked 'follow' on his NaviPad and is anxiously watching where she's going.

An unctuous salesman's voice has started murmuring 'comprehensively refitted for modern life, Bromley Reborn gives the lie to the sneer of 'slumburbia'! Your unique chance to relive the glory of the fin-de-siecle 'executive' lifestyle, with fully fifty percent higher Standard Space Entitlement'. Chang says 'wow'. Miriam loses patience following the meandering roadway, and zooms the virtual viewpoint straight through one of the houses. There's a brief glimpse of the extra insulated skin, a warren of little bedrooms, and the elaborate new eco-services cleverly fitted in to what would originally have been a double oilcar garage, then she's out through a cod-Gothic conservatory into what even this sales visualisation can't disguise is an arid open space, with wilting shrubs fringing a brown and dusty lawn. The picture freezes: she's pressed pause, and now looks up at Chang. 'Big garden too', he says hopefully.

'Wasteland, more like' she replies. 'Look, they can't store enough water to keep it from frazzling in summer, and we couldn't afford to buy any at summer rates. And we don't need that space entitlement. We get 30 square metres extra when the baby's born anyway, wherever we live. They have to give space entitlement away to try to persuade people to live there. And it's less than it seems anyway because you need loads of space for your own water storage, CHP, biowaste digestion, laundry and recycling stores, because the houses are too spread out to have them communal. And that means you've got to worry about keeping all the kit working yourself. And you can't have communal visitor rooms either because your visitors would have to walk miles down the street from them, so you've got to keep empty spare rooms if you ever want people to stay.'

Chang protests: 'but think of that executive lifestyle!' Miriam's scorn is withering: 'That used to mean jumping in an oilcar every time you wanted a loaf of bread. Now nobody can afford the carb for that, it means having to hang about till there's a pod that's going the right way – and that can take ages because everything's miles away in different directions. Unique? Pah, there's stuff like that outside every town. People abandoned it when the carb got too expensive, and now the government is paying the same developers who made all the mistakes first time round to come back and try to put them right. But whatever they do it'll still be Zeros housing for Zeros people' she said, using the sneering term now used to damn as *passé* everything pertaining to the first decade of the new Millennium, now seen as a nadir of bad-taste over-consumption. 'They probably still have those raucous parties with alcoholic drinks and bits of charred meat instead of nice gentle ones with dope and dal. Not the sort of lifestyle we want to bring our kid up in, is it?' 'No, perhaps not', Chang reluctantly agrees, flinching with distaste at the thought. 'OK, so what do *you* suggest?'

New Dunster

With three elegant flicks of her DigiThimble, Miriam zaps the Bromley brochure, pulls her own favourite back from under the mango plate, and sends a copy gliding across to Chang. It says 'New Dunster' at the top. Now Chang pilots the virtual viewpoint while Miriam anxiously follows. The picture shows three parallel wooded ridges snaking out from the edge of a town into open country, with small lakes and wetlands glinting between them. Puzzled, Chang asks, 'Very pretty, but where do people actually live?' 'Go in closer', Miriam softly replies. As he does so, Chang sees that the tree canopy is light and broken, and catches glimpses of big windows fringed with luxuriant vegetation. The sales commentary explains:

> It looks like a hillside with rooms excavated in it, but in fact it is a long apartment building, rising up to twelve storeys above ground, with a gently rising tiered southern flank, covered in earth, in which every dwelling merges seamlessly into its own private garden-balcony. Trees are an integral part of the architecture, moderating the climate, slowing the winds and provided shade in summer, while shedding their leaves to allow more light in winter. Digestion of the leaves and burning of trimmings contribute to the settlement's negative carb score. The protective vegetation and huge thermal mass means solar gain is never oppressive, but is enough to maintain a comfortable internal temperature all year round despite the negative carb performance.

There is space below the inhabited levels of the blocks for all standard modern eco-services. An internal tramline with frequent services will stop directly under your home. There are local shops, schools, and workspaces in the northern slopes of the blocks, with open outlook but not too much solar gain. Sheltered belts between the blocks are formal and informal gardens and recreation areas, market gardens and lakes and wetlands which provide waste water treatment, wildlife and recreation simultaneously, while incineration of the reeds also contributes to the development's negative carb rating.

'Okay, negative carb, we get the message, but there's more to life than carb', Chang protests. 'Of course there is, my love, but negative carb means we can have it'. 'Come again'? 'Look, negative carb doesn't just mean we wouldn't pay any Greenhouse Tax – though that's handy enough with the rates going up all the time. It means we could actually store up some carbon credit as we live day to day, instead of having to pay out. Then maybe we could have another midsummer chill-out in the Arctic Circle for our tenth anniversary. Or one year your dad could fly from Pakistan to see us instead of having to spend that whole week on the train.'

Trying not to be impressed, Chang says 'but didn't they try big blocks of flats in the middle of last century and then have to pull them down because everyone hated them and wanted to live in places like Bromley instead?' Miriam replies, 'Those were shoddily built, they were never maintained properly, they never had decent amenities, it took ages to get down to the ground and when you did it was just dull windswept grass covered in litter and dog poo. So of course people hated it, and never looked after anything, so they got neglected and broken down too. This is totally different. It's offering everything people used to have to go to places like Bromley for, but that *they* can't provide any more without the oilcars.'

Chang still looks hesitant. Miriam says, 'Look, love, we don't *have* to move anyway. We know we couldn't afford a bigger place round here even if one did come on the market. But just because we're entitled to more space with the baby, doesn't mean we've got to *have* it.' Chang's face brightens. 'It's nice round here. Nowhere else would be as tranquil and fresh as Brixton would it?' 'No, nor as good for schools and clinics and stuff'. 'And just a walk from the station when we do need to go anywhere beyond bike range, like my office in Manchester or yours in Brussels'. 'And it's not that bad for carb even though it's 160 years old thanks to the shared walls, the shutters, the new windows and the solar roof.' 'Now we've got the digital table we could give the digiwall away, and that would make more living space. Then you'll *have* to tidy your dirty plate away before you can watch the footie.' 'And it's nice to be able to say we live in Brixton.' 'Imagine telling people we lived in Bromley.' They both grimace, and then burst into fits of giggles. Their hands touch on the 'off' switch of the digital table.

Finding the ideal home

The point about Miriam and Chang is not that they are exotic and different, but the opposite. Just like people in 2011, they have ambitions to get the best housing they can. Just like us, their aspirations and preferences are formed from a mixture of 'objective' practical considerations and 'subjective' matters of self image and social status which, like now, are influenced by fashion, images and associations, and – let's admit it – snobbery and prejudice. Just like us, they would not be able to afford exactly what they might ideally want (even if they could agree what that would be!), and therefore have to make trade-offs and compromises. The story is concerned not with the utopian question of whether they living in their perfect dream house, but the more realistic – and interesting – one of whether they can get somewhere to live that meets all their basic needs, and that they feel generally comfortable and content with.

Lifestyle decisions

What is different is not the people, but both the practicalities and the image of different kinds of housing. The 'practical' starting point for the scenario is that some combination of taxes, permits and possibly rationing has made greenhouse gas emissions one of the most significant cost considerations in housing choices. The scenario mentions a Greenhouse Tax, with tax credits offered as in incentive for fitting low-carbon technologies, and implies that carbon intensive luxuries, including long-distance travel, would be somehow rationed so that people with low-carbon lifestyles would have more opportunities to do them. However, the details are left vague because they do not matter for the principle, which is that public policy interventions have deliberately made carbon matter much more in people's lifestyle decisions.

This has created immense market demand for a 'step change' in energy efficiency and other low-carbon approaches, ranging from simple familiar old measures of insulation and draught proofing to neighbourhood combined heat and power fuelled by burning or digesting biowaste, photovoltaic roofs, and the kind of managed, motorised control of ventilation and solar gain currently more familiar in expensive cars.

The scenario speculates that the car industry might prove better at meeting such demands than (at least parts of) the current construction industry. That would be a relief for those employed in cars, because the scenario also predicts that carbon taxation will also have made private motoring an occasional luxury rather than a daily routine. This in turn implies that housing built in locations, and at densities, which require occupants to drive to day to day amenities, will become inaccessible and inconvenient, while higher density urban housing nearer to shops, schools, public services, workplaces and so on will become much easier and cheaper to live a full life in. If, as the scenario posits, denser older housing will also prove much easier to retrofit to high energy and other resource efficiency standards than suburbia, practical considerations will reverse the current relative attractiveness of inner city and suburban housing: dense inner city neighbourhoods will be the sensible, convenient place for most people to live.

Importance of image

But what of image and association? The starting point for this scenario presupposes that, somehow or other, the public have become sufficiently convinced of the reality and gravity of climate change to make it politically possible for a government to take effective action. If and when this happens, it seems plausible that the lifestyles associated with the rejected complacent irresponsibility about climate change will fall out of fashion, and that the three-garage detached executive estate house will come to be regarded as naff as well as becoming inconvenient to live in. If so, these will be the new slums. Perceptions and realities would then be consistent and mutually supportive, unlike the present position where consuming more and more is seen as not only possible, but morally acceptable and a sign of success.

That is perhaps the most contentious aspect of this scenario. But the underlying message is independent of whether readers find plausible the particular speculations about images and perceptions offered. This message is that how people behave, and what they think and believe – about the reality of environmental limits and how we should respond to them, about what constitutes the 'good life' and the mix of private and public goods, personal freedoms and collective responsibilities that makes it up – will be just as significant for whether we make an effective transition to lower carbon living than technical matters.

This is illustrated by the reference in the scenario to another new policy instrument to manage housing demand, 'standard space entitlement'. The idea of rationing the amount of space people can occupy, or even taxing the excess above some bureaucratically decided norm, might strike people accustomed to the UK's deregulated housing market as shockingly, unthinkably, Stalinist. But does it really make sense that a country already seriously short of housing should treat the ability of wealthy homeowners to keep large amounts of habitable space empty for years as entirely a matter of sovereign personal choice?

14
The Refugee

Tanya Hawkes

My name is Nikora and this story is about my personal montage. The teacher asked us to all make a montage out of photos and images so that we can remember the past and look forward and plan the future. My montage is about my family, my journey from Aranuka to London and my plans to be a prime minister. Firstly, I wrote my name in the middle of the montage in large font: Nikora Ribetta. I put the English equivalent in smaller letters underneath: Nicholas, because even though most of my friends had English names, my Mum and Dad always called Nikora, even though I asked them not to.

My montage

To make the montage we sit and think for an hour about our life. This was quite funny as we all had to be very quiet and meditate on what we want when we are older. It's called the 70th birthday party – you have to imagine who is at your birthday party and why. All I could think about was how impossible it would be to fit 70 candles on a cake and that you'd probably need all your friends round to help you blow them out. My teacher said: 'that's interesting because that means you see yourself having lots of friends around you. You might want to put that into your montage.' So I did.

I also put a wife. I hope it will be Beth because she's my girlfriend, which means I can't ever show her my montage because then she'll know that I want to marry her. My mother said you shouldn't let girls know too much about what you want because then they'll lose interest. I can't help it though. Beth is one of those people that make me instantly happy. Once we were in the barn watching the kittens' play fighting when she jumped on me and tried to wrestle me. After five minutes of rolling around we lay still and she said 'I love you, Nikora'. It made my eyes go hot and watery and I couldn't say it back because my voice would have been croaky. I made her a little film instead, an animation of lots of cut out drawings. I reckoned we were equal then. It was hard making the film, but then I figured so is telling someone you love them for the first time, especially if they don't say I back because they might cry because it reminds them of their mum. There's no way I'd have been able to explain that.

The first part of the montage is the past. It's how you get here. I showed that to everyone in the end because Sarah, my teacher, said mine was very interesting. The picture shows an aeroplane. Apparently there were thousands in the air at any one time at the turn of the century. We came to Britain on an aeroplane from Aranuka. Aranuka is, I mean was, my island. It's hardly there anymore, nor are most of the other islands that made up Kiribati. The other people who used to live there mainly went to live in Australia or New Zealand. Except me and my Mum and Dad. We came to Britain.

When we arrived in London I was 8 and it was crazy. I put an image of myself at 10 years old as I didn't have any images of me when I was 8. I'm next to David, who was my host Dad. He's got one arm round me and the other round my real dad. They look so different. My dad looks small and is frowning. David is tall and smiling. I'm looking away from the camera, up at something in the sky.

David brought us to London. He worked with my Mum. They used to talk to each other through computers a lot and he came to stay with us twice. My Mum used to stay up very late reading documents about 'climate change' and 'forced migration', and then spend all day in boring meetings. It was all very important apparently.

`When we emigrated it was very sudden. David was staying with us after the Conference of the Parties in New Zealand. He came in said 'look, I can get you three on the plane with me, but we'll have to go in the morning.' That was 14 December 2022. I remember the date because it was Louisa's birthday. She was my best friend. I always used to measure her on her birthdays, because once when were little, she told me that birthdays always made her feel taller. She used to make me instantly happy too. I couldn't go to her party and we never said good-bye. I tried to send her messages, but I never heard from her again. My mum said that she probably ended up in Australia. I sometimes wonder how tall she is now.

I was crying on the plane – I've still got the photo of when we got to London, It was in the paper – Mum and Dad look tired and I look sad. David was angry – but not with us. He made a speech to lots of reporters about justice and how thousands of children were going to die. I thought about Louisa and whether she would die. I remember what David said at the end: 'Say goodbye to the policies of selfishness and greed and hello to the new era. The era of compassion and co-operation, welcome in the Good Society.'

There were cameras everywhere and afterwards a lady with long grey hair took me to one side and asked me lots of questions while a man filmed me with a camera: 'Was I scared? Did I enjoy the plane journey? What was my favourite book? Was I pleased to be in London?' I liked answering them until she asked if I was worried about violence from the Defend the English League. I didn't know who the Defend the English League were. I asked if they were the army, then my Dad and David stopped the interview and got angry with the woman.

I searched and found the article and put in on my montage. It's a bit embarrassing now. But I look quite happy in the film, until the bit where my Dad started shouting.

RefugeBritain

I found an image of a house that looked like the one we shared with David. The address was 36 Bakers Hill, Hackney, London. We were at the top. I remember walking up lots of stairs with prams and bicycles in the corridors. The flat was very big and there was a huge garden on the roof-full of fruit and vegetables. There were two pigs in a pen, and some chickens. David said roof top gardens were the best sort because the foxes couldn't get the chickens, which he said made him happy because he didn't want to kill the foxes, unless it was absolutely necessary. He often said those words 'absolutely necessary.' They were lovely days, helping David cut the fruit bushes and feed scraps to the pigs. I put a picture on my montage of some chilli peppers because I remember picking them off a bush and placing them on paper to dry on the window sill. I'd rubbed my eye half way through and my Mum had to pull me into the bathroom because I was screaming and she poured water in my eye to get the chilli out. I love chillies now though.

Mum, Dad and I all slept in the same room. There was one big bed which they slept in, the sheets all creamy white and the walls all orangey yellow, like a sunset. I had a little camp bed on the floor, but most nights I would wake up and then kiss my Mum gently on the cheek until she'd pull back the covers and let me under to cuddle her. Once I kissed her cheek and it was wet and salty with tears.

The best night ever in that flat was the time I woke really late and realised that Mum and Dad were still up and talking with David and lots of other people. They let me sit in the corner on the big comfy chair. I pretended to be asleep but I was listening to them and looking at the room through my eyelashes. There were five empty wine bottles on the table and lots of bits of left over food, which made me feel hungry. David was a bit angry with another woman, called Kate, but not very, because occasionally they would hug each other or grab each others hand. David said that he was resigning. He said the constituents didn't understand and that it was a matter of principle. Kate cried a bit and waived her hands about. She said that he was a coward – that the country needed people like him to keep going. That personal principles were no good – there were plenty of others who could keep up the refugee work. He would get arrested and go to prison and then he'd be no good to anyone. I couldn't stop looking at Kate – she had tumbling hair in many shades of light brown and her face changed all the time. One second it was sad, then cross, then smiley, like the pictures they made me use when I first came to this school and didn't want to talk: Sad face. Happy face. Confused face. Kate often came to David's flat and sometimes they would kiss each other. She kissed me once but only on my forehead though.

Later that night my mother told me quietly and tearfully that David was a politician. He was very close to the prime minister in the cabinet, which was funny because I imagined them sat in a cabinet, like the one we used to have at home full of lots of plates and cups. She said that he was a very good man who was making a point by letting us stay in his home. That because of him, thousands of people across Britain were helping people from other countries to escape from war and water shortages caused by climate change. They were called RefugeBritain. I liked their logo on the website: it was a small house containing a cartoon hedgehog. David laughed at it and said it was ridiculous logo, but he was glad I liked it. Britain was over its quota of refugees and the ones who were allowed into Britain mostly had to live in large prisons and huge ships. Sometimes families were separated into different bits of the camp and they were horrible. RefugeBritian helped the people who arrived in Britain illegally. They used to take buses and collect people, then drive them to safe houses.

She said that David might be arrested for letting them stay here and would be forced out of the government and that's why Kate, his girlfriend was so upset. I didn't put any of that in my montage. It was too difficult to make a picture out of, so I just put a newspaper picture of

David layered on top of a picture of their flat. The next bit of my montage was even more difficult. I cried as I drew a picture of the prison where they locked up my Mum and Dad.

Human Centred Climate Adaptation

After Mum and Dad were taken away, David persuaded the police that he was applying to adopt me, which surprised me because no-one had told me that before. Later I learned he was lying to the police. He said sometimes it's Ok to lie, because occasionally 'people behave like idiots.' He said I would never need to be adopted, because I have amazing parents, and he would do everything he could to get them and everyone else out. He got me a Carbon Card. I was too young to have one, but he did it anyway. He said to keep it with me all the time and he would make sure that it always had some credit on it so I could travel and buy a few things. Not much, because it would be out of his rations, but enough to get by.

David found me somewhere to live in a country called Wales – a little town called Machynlleth, which is where I am now. It would be safer for me in the country, he told me. He had friends there who would look after me and find me a school. He said he would miss me a lot and he never once regretted that he brought us here, but because of his job he had to do things he didn't always agree with and that he was very sorry about my mother and father, but that hopefully, soon there would be a different law in Europe which would make all the countries get rid of the horrible camps and ships and let people, *all* people – he emphasised that word – live the lives they deserved.

He gave me a document he had written, a paper copy, because at the time I didn't have phone or computer in which to store it. I cut out the title of it and scanned it onto my montage: *Common Futures: Human Centred Climate Adaptation Policy* and put the rest of the paper carefully back into the folder that had kept in safe for the last five years.

By now a third of the screen was filled with images, clashing against each other. Colours that didn't match. I couldn't put in any pictures of my life in Kirabas because none of the pictures looked like it, apart from one that had a picture of clear blue waters and bright sunshine. I placed it underneath the plane.

In the future section of my montage I put a picture of a family having dinner: a mum, a dad and a teenage boy. I looked through pictures of politicians, but I couldn't find any that represented my future as prime minister. In the end I found an image of David just before the end. I didn't really want to remember him like that, but it was the only picture I could find to show what I would do if I was the prime minister. I put three words over the top of David's image: 'The Good Society.'

Thirty years later

Nikora is on a train tapping onto a screen in front of him. Kate, his senior aide, is reading to him from a screen in front of her. They are heading towards Paris, where the UNFCA (United Nations Framework for Climate Adaptation) are being held.

> 'So ... Mister hot shot shadow minister ... what are you going to say to the world, when you arrive at the summit? Let's go through your lines so you don't mess it up like you did in Baku.' Kate waited for Nikora to react.
>
> 'It wasn't my fault. I thought Baku was the country, not the city. Azerbaijan. See I remember its capital now! You didn't brief me enough.'
>
> 'Aaaa- sure- *bay*- shaaaan. You still can't pronounce it properly.'
>
> 'Hardly matters, does it? It was already a dessert. Don't worry, I think I know how to pronounce Paris and last time I heard it was the capital of France? Go on then; throw your worst at me', said Nikora.
>
> 'You need a story ... The citizens of Britain have stories. You're going to need to know their stories. Like Martin who lives in Edinburgh – he sent a message to the office the other day. Care to watch?' Nikora nodded. She turned the screen towards him.

The Good Society

Martin's angry face popped up on the screen. A small girl sat next to him on a sofa:

> I'm Martin. I'm 37. I live in Edinburgh. I want to know what your party is going to do if you win the election. Your government, eight years ago, put me out of work. I'm a solar water heating installer – I remember all the promises – thousands of jobs, retrain for this and that, wind and solar is the future. Well that was all great, but no-one installs solar anymore – there's hardly any new buildings – the booms over. Now what am I supposed to do – tell me that. I've asked all the parties the same question. Now that most of our energy comes from abroad – what are we supposed to do here in Britain? Twiddle our thumbs? One million climate jobs! It was a con. How are you going to help people like us if you get elected this time?

'Don't roll your eyes', said Kate. 'This is just a sample. You're out of touch and you've got to get with it. This government is going to bring back coal production and end carbon rationing and God's knows what else, unless we can come up with an alternative. People like Martin here' she waived her hand at the screen, 'are pissed off. They can remember what it was like when we had oil and coal – we were rich. Return on energy investment was higher than anything you can get with wind and solar. You know it! Everyone wants the good old days back. What's you're line going to be?'

'It's easy! We tell the truth. We represent the Good Society, remember 'Bigger than self' policies. This government won by a thread – and they'll lose next year, you know it!'

'But this is popular. This will win them the election again. It's genius. Us, the little Britain's, against the rest of the world, trying to regain some power so we can pull ourselves out of poverty with greater energy production. Export our 'clean' coal to the rest of the world – we've got loads of it – remember ... 'Britain led us into the industrial revolution in the 1800's now we can lead the world there again, but safely this time.' *'That's* going to be their headline.' Kate was tapping on the train window, 'Look out there. It's the same as London: people trudging around the fields growing food, brown-outs during the evenings, squatter camps on the edge of every major city. Hundreds of thousands of people migrating.'

'It's nonsense. You know it. They know it.' Nikora stretched himself up and ran his fingers quickly through his hair 'Because the alternative is worse. That's food to feed people, in those fields. Squatter camps instead of refugee prisons; freedom to move to different countries. Better that than the morally redundant society we had before. We never invested enough in other energy – we can just say we'll invest more – everyone will become richer.

What investment? Where's the money for that going to come from? We print it! Like we've always done.

Yes, but you can't *say* that – people don't like to hear about debt. They get scared.

Oh come on – people aren't stupid. Britain's been at the so called brink of recession scores of times. It never used to stop the growth in arms and financial services. Let's grow some renewables. Let's grow some more public services.

'You don't need to convince me. Convince the citizens.'

Nikora leaned back. 'Why are we sat in first class, anyway? What's that going to look like when we arrive?' Kate laughed. 'Don't change the subject. You politicians. You're all the same. If we were in the second class carriages we'd have no room to breathe, let alone get your political career back on track.'

'Back on track. Is that some kind of train joke?'

'What made you become a politician, Nikora?' Kate leaned in to him. 'Have you forgotten what you went through? What your parents went through? What drove you here? I remember sitting in David's house with your mum and dad- the fear of waiting for the knock on the door. David, giving up everything to help them.' She looked away for moment, moving her hair so it covered her eyes. 'Tell them that – that's the story they need to hear. It's all about stories. People need their heroes. Be their hero.'

'Ok, I get it.' Nikora shut his eyes for a minute to stop Kate from talking. She'd gone quiet, anyway. She always did whenever David's name was raised. He reached into his bag and pulled out the comforting old document, the title cut out for his personal montage all those years ago. It was worn and yellow. The paper was thin at the edges. David once said that paper would survive everything. 'If something is really important, write it on paper – it separates it out from all the other junk.' There were still penned notes in the margins in David's generous handwriting – words like 'Really?' and 'Yawn!' Fun words, as if he found political jargon tedious. Kate looked over. She saw the document and then touched his arm.

'You miss him, don't you? Me too. You don't have to do what he did. He wouldn't have wanted you to.' Kate turned back to the window.'

What would David do?

He wouldn't have cared about a meaningless career, thought Nikora. He would have cared about the Good Society.

Building the links

The train sped past the endless rows of poly-tunnels and gave way to the makeshift buildings of the squatter camps. People called them slums: *dilapidated neighbourhoods where people live in a state of poverty*, according to the dictionary. But those people had never walked around them or lived there, thought Nikora.

Whenever he went to stay with his Mum and Dad he was amazed at the tiny schools on street corners, the shops that sold little hot potatoes for commuters on their way to work, the elected spokespeople on each street who went and represented their neighbourhoods at council meetings. Small urban communities all jammed together in rows of temporary and semi permanent houses. It was a miracle of human existence – creative, chaotic. Mum and Dad lived on one of the boats – there were hundreds of them, bobbing against each other in the canal – little, painted toy houses. If you took a photo from the air it would look like the montages they used to make at school. You just kept adding bits to it. No planners or architects, just people from everywhere buildings extra walls, rigging up electricity cables. In one part the markets stall covered the tram lines. When a tram came through, the stall holders would just move the stalls out of the trams way and then slide them back over the tracks once the train passed. Every street was a slice of a different country. One moment you were in Poland, passing tiny vodka bars, the next you were in·Morocco, following the smell of cinnamon and coffee.

The National Assembly parliament building seemed boring and outdated in comparison. All colonial grandeur, the pillars like huge barriers, keeping the people out of politics. They pulled up at the same time as lots of other official heads of state in their official cars. Kate manoeuvred Nikora across the road.

'Stand in front of the statues – they'll fit with your speech.'

'It's going to be a long one.' Kate sighed. 'If you must. It's your funeral'

Reporters gathered. Nikora's face projected onto the big screen outside the parliament building. Everyone wanted to hear about the leader of the opposition – the one who was insisting that the UN energy inspectors enter his own country and then impose sanctions if necessary. The one who was about to commit political suicide.

'Why are you calling for the UN to intervene in Britain's energy plans? You'll be the least popular MP in Britain won't you?'

'Quite possibly. But being in politics is all about the tough, unpopular decisions, isn't it?

Will you be meeting with the prime minister while you are here? They'll be no chance of you forming a coalition now will there?

Probably not, no.

More questions. The Adaptation talk must have got really boring, though Nikora.

'Listen,' he started. 'Once we divided up all the political issues into different sectors. Hard to believe now, but we had separate departments. The Department for Energy and Climate Change, the Department for Transport the Government Advisory for the Purchase of Wine ... (*Laughter*) no really – it existed. Apparently it was the most productive. (*More laughter.*)

Anyway each department had its own staff and very rarely did the policies form each department join up properly. So the Transport Department would suggest a decade of road building while the Climate Change Committee advised on deceasing carbon budgets. It was chaotic and contradictory.

And people issues ... it was as if people lived on another planet.

The climate issue changed that; but not at first. We had separate departments for transport, food, housing, and environment. People put everything in political boxes. A culture of ticking off each goal regardless of whether it was useful and fitted with other policies. No-one was prepared to admit they were wrong. You might be too young to remember when the Environmental Audit Committee advised the Government to embed climate policy into its highest office. Nearly 50 years ago, that turning point changed everything. We forget now how important that was. When the government then, started the ball rolling to introduce Tradable Energy Quotas they had no idea that they were introducing much more than a system of personal carbon trading. It was ironic really – the Conservative government at the time ended up introducing the nearest thing to Socialism for decades. But that's another story.

The unexpected side effect was phenomenal. Everyone began to see the links: the link between energy and economic growth. Fossil fuels allowed us all an unnatural return on energy investment that caused unprecedented growth in a way that has never been replicated since. Everything slowed down, once we rationed carbon. It was as if everyone could see the effect they were having on others around them – as if a veil had been lifted from our eyes. I remember in our street and elderly couple ran out of carbon credits and everyone clubbed together and donated some of their own to se them through the week. That rarely used to happen when we just had money.

And more importantly, much more, we started to see what was happening in other countries. The whole political and economic arena shifted when we realised we had to count all our own carbon emissions – we could no longer consume products made in China and then make them count the carbon cost. Revelation after revelation went on for years as the link between the things; all those *things* and the energy it took to produce them became obvious.

And we mourned. I know it was difficult for people. Giving up the lifestyles we'd got used to was hard. But they were starting to unravel anyway – we know that now. It was all an illusion that couldn't last forever.

Things changed. They changed for the better. They changed for the better for everyone. And that is the point. That is why I can no longer stand by while the government of Britain reverts to producing the most dangerous substance in the world. That is why I am urging the UN to impose sanctions on my OWN country. There will be those who say I am a traitor. There will be those who will put me in prison. My career as a politician will be over. I don't care about that. This is what I care about.

Fortress Europe

Some of you will remember David Hawkins. For 20 years, David worked tirelessly to make this world a fairer place. He risked his own freedom. Even if you don't remember him, you will probably remember the phrase he was famous for: the Good Society.

David Hawkins dreamed of a society that catered for everyone. That recognised that if one person failed, then we all failed. That's why he put everything in his life at risk to help me and my parents. My parents, wonderful people who I owe a great debt to, spent nearly five years in prison. A prison that we used to lock up refugees; hundreds of thousands of them. We had created a low-carbon society. We had given people an equal right to carbon in Britain, but it took us another two decades for the phoney economic borders to crumble and for people to be able to move freely from country to country. It took decades to reverse the culture of regionalisation and securitisation. 'Fortress Europe.' That's what David called it. We were all locked up in our big prison, scared of running out of energy, scared of running out of food and water, scared of people from other countries. But our prisons were all in our mind.

This beautiful National Assembly building is a testament to the triumph of democracy over autocracy. On the one hand we have the parliament with all the liberty it represents; behind us we have these statues of African slaves to remind us that there is rarely economic growth without exploitation. That struggle continues now and always. The struggle to create a Good Society, not a system where 'a huge community of producers deprive each other of the fruits of their collective labour' – but a socially just economy, 'accompanied by an educational system that would be oriented toward social goals.' (That was Einstein, by the way, in case you were wondering).

We are told our current Government in Britain has the will of the people. They tell us this policy is merely about creating more energy. That it carries no risk to the climate. Well, it may well carry limited risk to the climate, but why even take limited risks – once we open the coal face again, countries all over the world will begin to burn the most polluting substance on the planet. Do we trust countries to really be able to separate out the cleaner coal from the dirty power plants? Why would we want to return to that risk and uncertainty?

But this policy is not just about creating more energy. The knock on effect of unlimited energy production is the psychological effect of false abundance of resources. It's an illusion.

When David Hawkins was jailed for leaking government documents – the ones that showed us how millions of refugees would be left to die, shut out of fortress Britain at worst perishing on the borders of Europe, at best imprisoned in squalid detention centres – he reminded everyone of our collective humanity. That not only should we have energy equity in Britain, but that we need to extend that equity to whoever else needs it, wherever they come from, because that what a Good Society does.

Eventually the Good Society failed David. Just after his tragic suicide in prison, he was issued a pardon because so many people across Britain and the rest of the world called for his release. He never got to see the millions of emails, letters, websites that were dedicated to him and his work. Two year later the detainment of refugees was ended and economic borders became history.

Let's not undo that history now.

I call today for the UN to intervene on behalf of the British people and stop our government from plunging us back into an energy system based on fossil fuels and all the economic and social dangers that go with it.'

'Even if it means economic sanctions on your own country?'

'Yes. Even that.'

'Nikora moved away and Kate took over. He needs to go into the meeting now. You can ask me any questions from now on.'

Nikora looked back at Kate fielding the question with such political ease. He'd have to face the prime minister later who would no doubt be plotting Nikora's downfall. He wondered what it would be like to disappear into the squatter camps and never be seen again.

15

The Gun and the Sun

Horace Herring

I woke up and glanced at the clock: 7:58 it blinked. 'Lights on,' I mumbled sleepily. The ceiling lights slowly increased to a dim glow that did little to relieve the gloom in the room. 'James,' I shouted,' More light.' His smooth voice instantly replied, 'Sorry sir, no more light. Power level status is critical'. I turned over and looked blearily at the large wall screen, his face was absent. I sat up, and the chill in the room struck me. Then I remembered, we were in the dead days of winter, and today was the fourth day.

Another dead day

I leapt out of bed and hurriedly dressed. I pulled back the curtains but couldn't see much beyond the garden fence. 'James, what's the forecast?' I cried despairingly. 'Like yesterday, cold, foggy, no wind' and then he added, 'Generation prospects zero.' Another dead day, I thought, the once in ten years event, we had so blithely dismissed at the tenants' meeting last June. Why invest in more storage for such a rare event, we all agreed on that blazingly sunny day. Instead we voted to upgrade the batteries for our communal cars.

'James,' I barked angrily, 'how much longer till a cup of tea.' He was silent for a long time, and then replied somewhat despairingly 'I'm sorry sir, I can't work it out, the kettle is old, I don't have its heat loss parameters, I don't think it will boil today.' I was surprised by his emotion; perhaps it was really possible to upset a robo-butler by giving it an impossible question to answer. 'Ok, ok, forget it,' I replied somewhat contritely 'I'll go to the Café'. 'Oh yes sir that would be best,' he instantly responded, with what seemed great relief in his voice, 'they have special offers today on breakfast: porridge, pancakes, muffins, and toast ...' I cut him off in full flow, and wondered if he needed reprogramming. He seemed to be getting too emotional, but then wondered if the fault lay with me for being too unemotional. I was desperate for a cup of tea, but the thought of going to the Café and seeing all those people, who would talk to me, filled me with further gloom.

The room suddenly brightened and a face appeared on the screen, the Mayor's face. 'Attention, fellow villagers', he began in his usual pompous manner.' Today is the fourth dead day and power supplies are critical. To conserve supplies we are cutting off electricity to all non-essential users. All villagers are requested to vacate their homes and assemble in the Village Hall. All planned work assignments are suspended; await details of future assignments.'

He paused and then tried to look more concerned, 'Rumours have been posted about the Convoy being camped nearby. I must stress they pose no threat to our safety; our defences are excellent with a proven 90 per cent success rate. The Sheriff has the situation well in hand. And negotiations', here he pulled a face, 'are underway to facilitate their departure. There is no cause for any alarm.' His face faded from the screen and the lights almost went out. They would totally in ten minutes.

I stared out of the window into the whiteness with mixed emotions. At first relief, that the meeting of the Local Food Committee had been cancelled. Three hours of total boredom as we discussed for the umpteenth time whether it was better to import organic oranges from Spain or go for the local genetically modified ones. Did we really have nothing better to do with our time! As I gazed out, I was filled with gratitude that I was not one of the ferals in the Convoy, living out in the woods in this freezing fog with no power: a desperate situation. Then alarm at the thought that ten per cent could get through the fence; sure 90 per cent would be disabled – our lovely euphemism for killed – but what of the ones that did? If your chance of survival in this weather was only 50:50 perhaps a ten per cent chance of life seemed an attractive option, especially if it was not your life at risk but some hapless fighter who you hired for this sort of work. And then I realised the significance of the Mayor's talk of 'negotiations'. Basically it was all about energy, and in particular biofuel. How much did they need to get their motley collection of vehicles going again, and what would we be prepared to pay to see them leave. At heart, a rational cost-benefit problem, which our War Games programme would be currently working on. A question of probabilities: on the weather, on their desperation, on our defences.

If ten ferals got through, they could destroy our energy storage systems, and cripple our community. It had happened elsewhere. With no biofuels, no tractors, no planting, no crops. ... With no batteries or flywheels, no electricity storage and we would be totally at the mercy of the vagaries of the sun and the wind. There was no grid to connect us to alternative sources. That had been long destroyed since the Troubles, and now rural England was a patchwork of small self-sufficient communities, either static or mobile. When the sun shone, and energy was cheap, as in the summer both sides rubbed along, benefiting from barter and trade in a whole host of goods and services. When it didn't, like now, they struggled to survive and we paid the consequences.

The lights went out completely, and I wondered what I should do. The Village Hall would have all the false joviality of a cocktail party; everybody talking too loudly and pretending it was all great fun. The long queues for the toilet, the monotonous stews, the snoring at night, and the smell of hundreds of bodies cramped together. And worst of all would be the conversations. It always turned late at night to the Troubles; an agonised introspection without answer. Whose fault was it: the greens, the browns, the reds, the Europeans, the Islamists? I was totally sick and tired of discussing it – what happened had happened, and there was nothing we could do to change it. And as for the future, I had given up hope. All I wanted was to be left alone, to just sit in the sun, drink some decent coffee and read a good book. And here was as good as anywhere else. If I could just coast along without any tiresome involvement with others I would be happy.

I stopped suddenly, and looked about guiltily. I had just committed the most grievous offence of 'freeloadism' – wanting to benefit from the community without being willing to contribute. This was an expellable crime in some communities, but luckily ours was very tolerant. If my thoughts were known, I would have to have counselling with the Vicar, and take part in the community-bonding rehabilitation programme. I was filled with horror and tried to think back over my recent behaviour. Had I exhibited any telltale signs of my disenchantment?

The Wellsians

As if reading my thoughts, the screen burst into life and the Mayor's face appeared, 'Simon, not gone to join the others?' he said in what I took to be an accusing manner. I stammered, felt hopeless and fell silent waiting my fate. 'Simon, something's come up. A task just for you, if you're available.' He paused and I looked up hopefully, perhaps I hadn't yet been spotted. 'Yes, Mayor', I said brightly, 'Any way I can serve'. He looked at me suspiciously but went on 'A party of Wellsians has just

arrived, and they want a tour of the community. They specifically asked
for you to be their guide. They will be here until the fog lifts.'

I was totally amazed, my mouth hung open, and I stammered
'Wellsians, here, in this weather ...' He shrugged and said off-handedly,
'Well, you know what they're like ...' Yes we all knew the Wellsians:
crazy scientists based at Sizewell who preached and lived the life of the
unlimited benefits of unrestricted science and technology. No scheme
or vision was too big for them: nuclear power, geo-engineering, weather
control, colonies on Mars, human genetic modification ... basically
everything that this community was against. Once I was too, but now
I was tired and I found their boundless enthusiasm refreshing. They
were good to argue with after years of stale internal conflict over petty
matters.

I must have fallen into a daydream, as the Mayor's voice rasped,
'Simon, are you with us! This visit is very important. The work credits
are maximum, and you can use the visitor's office during their stay.'
I instantly came alive, the visitor's office had a small couch in it, I could
sleep there at night, and their visit was a perfect excuse to break away
from any tedious conversation I might encounter at the Village Hall.
I replied, more full of life than for many months, 'Yes, Mayor, I shall be
delighted to undertake this work assignment'.

An hour later I was at the Visitors Centre greeting my party. Twelve
of them in their neat uniforms, though I noticed cuts and bruises on
some of their faces. I had just time to scan through their bios; ten space
cadets, one space engineer and one Frenchman, no rank given though
the name somehow seemed familiar. I groaned, the Wellsians were arro-
gant enough, but to have a French one telling us how wonderful life was
in France! We all knew it was, the exemplar low-carbon nation, rich in
nuclear, wind and solar energy; a leading exporter of green electricity
and hydrogen throughout Europe. But we weren't such gritted-teeth
admirers of their foreign policy in North Africa. Solar was now the new
oil, and the French Euro Legion had been engaged for years in a brutal
suppression of the Maghred Islamists who wanted greater royalties from
the Desertec grid that bought such wealth to France.

The Wellsians were at first the ideal tour group: polite, attentive, and
well informed but I knew that soon their questions would become more
critical and slide into open sarcasm. This was not a good time to be
advertising the benefits of a solar-powered community. I kept on wondering why they were here, was it out of choice or desperation? Were we
the only refuge they could find? There was little love, and much hatred
between them and the ferals, who they treated as barely human and

just a source of genetic variety in their eugenic quest to breed a better human being.

It was a raw day, and I noticed some of the Wellsians shivering, they didn't seem to have any coats. But they all looked happy enough, apart from the French guy who looked miserable. We had passed all the things that I was happy to show off, our Village Hall, the school, the café, the shop and the office complex all still working thanks to our wood chip CHP plant. Now we were passing down the High Street, not a light in any of the house, all dead. This is where, I knew, things would get difficult.

The grid

'Bet you wish you had some of our nuclear power, eh' shouted a voice at the rear. They nearly all laughed maliciously. I chose my words carefully, mindful that our conversations were being picked up by nearby monitor. 'Yes, at a time like this, we would welcome a grid connection but as you well know there are some people who won't allow this' and I waved my arm in the direction of the woods. The Wellsians fell silent for that was their greatest obstacle, sitting there with a one GW power station, unable to export anything. Grids were too vulnerable to attack, to sabotage, and so expensive to defend.

There was only one left in Britain – the Great Green Corridor that ran along the route of the old M1. It was five miles wide and heavily defended with robo-guns and drones. The life expectancy of anyone entering it without authorisation was given as 43 seconds. In it ran the Euro Supergrid, transferring Scottish wind power to London and the few surviving English city states, and beyond to Europe. It also contained the hydrogen pipeline, the freight roadway and ultra-high speed train line. It also served as a haven for wildlife, as species slowly migrated northwards. In contrast they Wellsians existed on isolated coastal sites, cut off by geography and their extreme views. They had power, too cheap to meter, but what could they do with it except power some industrial labs and their highly popular City of Light, a giant amusement park and spa. Oh, and a free top-up of your car batteries whenever you visited.

Sizewell was now more of a theme park than a generating plant and it irked them greatly. For us, the future was decentralised, living on what we could produce, and we were a lucky community with a forest nearby. The sun might not shine, the wind might not blow, but the wood chips were always there, and we could trade forest products for biofuels. Yes our life was far more seasonal than in the fossil-fuel days – winters were very restrictive but in the summer we could travel freely. The Sunny Chef chain even offered, some days, a free top up with a cup of coffee! Except of course abroad. Since the Troubles, a time that still haunts the elder generation. It was a time of collective madness, destruction and turmoil that left us a truly broken Britain that no one could put back together again. I must have fallen into one of my daydreams, because the French guy was shaking my arm and saying, 'Sir, are you all right ...' The others just looked contemptuous. 'Oh, yes, I'm fine, sorry just thinking about ...' I suddenly remembered where I was and stopped. The Vicar took a dim view of people who dwelt too much on the Troubles.

I moved on down the High Street, talking about our low-energy homes. One of them interrupted, 'That's only because you restrict the appliances they can have – no toasters!' The others laughed. 'Yes, we have a power allowance for every home, 500 Watts for a single person, 1000 Watts for a family, so that does limit the use of some appliances but if I want toast I can go to the Café. All our needs are meet here, everything we need for a good life is provided.' I saw we were still near a monitor, so I decided to lay on our green PR, 'Yes this community fully meets the Round Tree standards, meeting all physical, social and sexual needs. We have been runner-up twice for the Green Community of the Year Competition, and there is a long waiting list of people who want to join us. We in New Corby are a very popular and successful community. It is in times like this, difficult times, during the dead days, that our community shows its best side.'

I glared at them but their faces were blank, except for the French guy who was faintly smiling. As we walked on over the village green towards the farm workshops, the French guy fell in besides me and extended his hand, a rare thing for a Wellsian to do as they generally shunned all physical contact. Greatly surprised, I shook it, 'Jean-Paul', he said with a faint French accent, 'My mother was English but I grew up on Belle-Ile'. And then I remembered, the name, the English mother, the island.

Hostiles approaching

But before I could say any more we heard a distant burst of gunfire. A nearby loudspeaker announced, 'Hostiles approaching, Farm Lane sector.' There was more gunfire, it seemed to be getting closer, and then an explosion muffed by the fog. One of the Wellsians announced calmly, 'We'll wait here until it's over' and they sat down on some picnic benches and started to talk quietly. Jean-Paul looked around anxiously, perhaps searching for somewhere to hide. 'They want me,' he implored, 'They knew the French will pay ransom, my father he's rich, he's ...' I grabbed him by the arm and lead him to a dead zone, where we could be still be seen but not overheard. I always identify such places before I take a tour party out; it saves a lot of embarrassment with the Mayor later.

I pretended to be showing him the herb garden, as the words tumbled out of him, 'In France we not like these people' and he gestured at the group now smoking some sort of cigarette, 'They're all mad, crazy, suicidal.' The gunfire was becoming more intense, and the loudspeaker announced in its flat voice, 'Security fence breached'. Just a few hundred yards away. I looked over at the Wellsians, they were now standing up and passing round a small flask, some were laughing. They had the air of people waiting for a football match to start. The smoke from their cigarette rose vertically into the fog. I thought of telling them that this was a no-smoking community, but then I thought what was the point, we might all be dead in five minutes.

It was all an act, exhibiting the legendary Wellsian stoicism and bravado in the face of death, and it was all being recorded. I glanced at Jean-Paul, his eyes were closed, his lips were moving and he was clutching a medallion that was on a chain around his neck. A Wellsian, praying! How they would mock him if they knew! There were several explosions seemingly nearby, and I was sure I could smell burning biofuel. Suddenly I was concerned about my status, my popularity in the community, how they would remember me. I know it was very low at the moment; I couldn't hide my indifference and contempt for their squabbles on the local food policy committee. We had spent four months discussing whether coffee should be banned!

If the Convoy overran the community, and ransacked it, it would be worldwide news: another sign of the barbarism still existing in England. And I would be on it. I turned towards the camera and smiled, and put my hand on Jean-Paul's arm and slowly moved him on, as if this was a perfectly peaceful sunny day. As we approached the Wellsians, I noticed their smoke was drifting away at an angle. I looked at the trees, the few leaves remaining were moving slightly. I glanced up at the sky, a pale orb showed. Then I heard the sound of the drones and I knew the Sheriff had arrived. The fog suddenly lifted and I could see the blades on our largest wind turbine slowly turning.

The battle I knew was now over, our energy defences were being replenished, while theirs were nearly exhausted after ten minutes of attack. It would be a massacre, a total rout. I felt sorry for all their fighters who were going to die, but then I knew I was being sentimental, they existed solely to fight and die. We rejoined the Wellsians, who were stubbing out their cigarettes and putting the butts in a small metal case. One remarked 'The Sheriff's got a good weather forecaster'. The oldest one, the engineer, smiled and I knew why they were here.

This occasion, I knew called for some special words, so I faced and camera and said, 'Gentleman, I am sorry for this rude interruption, shall we continue the tour.' They all laughed and a few cheered. We headed down towards the workshops. I could feel the sun on my face and a breeze in my hair. My newfound French friend was walking besides me, telling me excitedly about a similar miracle that happened on his island, and about the shrine there and how I should visit it. The gunfire faded away as the sun grew brighter, and I felt happier and more secure than I had for a very long time. I was sure that in due course he would invite me to make the pilgrimage. I could see myself sitting in a small French café by the sea, drinking an espresso and reading a good book.

16
Conclusion

Horace Herring

Once upon a time a rich man went to a lecture, and he was very impressed by what he heard. After the speaker had finished he asked a question, 'I'm a rich man, does your virtuous society apply to me'. The answer was 'Only if you give away all your possessions.' The rich man left deeply disappointed. From this tale came the proverb 'it is easier for a camel to pass through the eye of a needle than a rich man to enter the kingdom of heaven'.[1]

Throughout the ages we have heard similar message, from all religions, of the dangers of greed, gluttony and acquisitiveness (Michaelis, 2006). Now days we hear a similar lecture, about how we can only overcome the evils of carbon consumption by giving up our prize possessions – our cars, our steaks and our foreign holidays. In the cause of ecological virtue the rich are urged to lead a simpler life.

But throughout history few people have voluntarily embraced poverty or simplicity. We have generally sought to obtain virtue by toning down ethical texts, so that we only give away a little (such as with alms, charity or carbon taxes) or seeking to mitigate our 'sins' through compensatory measures (such as indulgences, donations and carbon offsets). The problem we (in the West) are seeking to solve is how to be rich and green: can we be ethical consumers? Can we still consume and live a low-carbon life? Or can we reinterpret the problem, as not one of lifestyles but of technology? Or maybe by even rewriting the text, so that 'greed is good' or 'economic growth is the solution to climate change'. The rich man then becomes the hero rather than the villain.[2]

Is consumption a problem?

In Chapter 4 we saw that we could have a 'decent' standard of life and still reduce our carbon emissions by over a third. However this lifestyle, with no car or foreign holidays, would be unacceptable to the many people who see the car as an absolute necessity and air travel as a well-deserved luxury. We may all be willing, as the government urges, to do 'our bit' but sadly the end result of everybody doing a little, is little gets done (Marshall, 2007). Thus major changes in consumption require major changes in lifestyle, but even extreme lifestyle changes are unlikely to reduce emissions by the required 80 per cent (Harper, 2007).

We do not live as our grandparents did, neither will our grandchildren live as we do. Tastes and fashions can change radically over time, what can seem a desirable lifestyle (e.g. the 1950s suburban family) to one generation can appear horribly archaic and undesirable a generation or two later (a point well made in the story in Chapter 13). We consume differently over time but not less. Our consumption reflects our incomes, our tastes and the technical possibilities. Unless our incomes are drastically reduced, we will not consume less. A long recession with falling real incomes, could make people shift towards lower-carbon (or green) consumption, but only at the cost of associating such consumption with 'poverty' and 'hard times'. Changing tastes (or attitudes) is largely ineffective if we cannot afford the green goods – what use are electric cars or solar PV if the majority of the people cannot afford them? So the best option is marrying green tastes with technical possibilities: making the green lifestyle not only attractive and desirable but also affordable. That is creating a mass-market for low-carbon technologies. At the moment there is only a niche market: the problem is how to expand it.

The technical options

Achieving a low-carbon society through lifestyle changes is not feasible, except if there were drastic changes in our society, such as its collapse (explored in the story in Chapter 15). Very few people will voluntarily embrace 'poverty', most want a stable, if not growing standard of living – which inevitably involves more consumption. Yes, there will be shifts in consumption towards more services and higher-value goods (arts, crafts, handmade items etc) but this does not mean lower emissions (globally), for labour has a carbon cost as well. The richer you are the more emissions you cause, and the quest for a low-consumption affluent lifestyle may be an impossible dream.

However a low-carbon affluent lifestyle is possible, if we could replace all fossil fuel use with low-carbon energy sources such as renewable or nuclear. There are many publications on this theme, some of them showing that it is technically possible to supply nearly all of Europe's electricity, and most of its energy, through renewables by 2050 (e.g. Czisch, 2010, WWF, 2011). For the UK, extensive modelling shows the feasibility of a low-carbon society by 2050, at an affordable cost (Skea et al., 2010). However this will require:

> Some combination of three core policy elements – carbon and energy pricing, technology support, and lifestyle and behaviour change. ... The political challenge is both to help to create those conditions and articulate a policy approach that is sufficiently ambitious and can command adequate public support.
>
> (p. 364)

It is easier to outline the technical measures that are needed than to outline the social and political policies required. The technical measures are already being put in place, both on the large scale – the North Sea wind farms and the European grid network – and the small scale – the biomass-fired district heating schemes and solar PV houses. However what are less developed are the financial and social policies: so far no carbon taxes on consumers and only limited subsidies (now via the feed-in tariff) for domestic renewables. Small-scale lifestyle and behavioural change is encouraged but not mandated, and is left by government to small NGOs and charities to implement (e.g. Church, 2011; Cambridge Carbon Footprint).

This lack of action reflects political realities; voters dislike anything that smacks of higher taxes or restrictions on their freedom to consume. This is well illustrated in Chapter 11 where there was public hostility, in Switzerland, to the proposals to place limits on the carbon emissions of cars (thereby preventing the sale of the largest cars) and to raise taxes on petrol. What the public did accept were technical measures, like building insulation or solar panels, accompanied by a system of subsidies and grants. The public overlooks that these subsidies and grants are funded by either general taxation or higher electricity prices.

Thus the technical options chosen will be largely dictated by public opinion, even though they may not be either the most-cost effective or equitable solutions In the UK, this means off-shore rather than cheaper on-shore wind farms (due to public opposition in rural areas), and generous subsidies through the feed-in tariff for domestic renewables (which benefits those rich enough to afford the capital cost). This 'painless' (or populist) approach to implementing a low-carbon society will only work as long as it is largely ineffective. Once it becomes widespread, the subsidies become intolerable and are sharply reduced by the government (as has happened in Spain, Germany, France and the UK for large scale PV), which can lead to collapse of the renewable industry.

It is pointless to pretend that the road to a low-carbon society is without pain or cost. Massive investment is needed either funded through general government taxation or higher energy prices. But the costs are bearable as they are spread over a long time[3], and consumers have (grudgingly) accepted a more than doubling in domestic energy prices in the last decade.[4] The main problem is a political will (driven by the lack of public urgency) on climate change: it is a far off problem with uncertain consequences. It is now (mid-2011) of less concern than our current economic problems (Shields, 2010). So what will drive the quest for a low-carbon society will be events, and they will have too be of sufficient scale to challenge our current energy lifestyles.

Challenging events

As Chapter 2 shows there are three forces driving us towards a low-carbon future – the threat of climate change, peak oil and energy security. None of these, so far, has given the public a compelling reason to accept unpopular measures. No natural disaster can yet be unequivocally blamed on climate change, oil prices at $100 a barrel are still acceptable, and no conflicts disrupt the smooth flow of global energy supplies. Each of these three are a possibility, but judged (by the public) as too uncertain to be worthy of significant effort or money. However, one recent event has had a significant impact: the earthquake and tsunami in Japan on 11 March 2011, which destroyed the Fukushima nuclear plant. This has influenced global nuclear policies; Germany has since decided to phase-out nuclear power by 2022 and switch to renewables (but most probably to gas, or even coal). However, if gas substitutes for nuclear, carbon emissions will rise.

What is really needed is some event that will lead to the phase out of fossil fuels, especially coal. High oil prices obviously encourage the development of alternative energy sources, mainly renewables but also of coal, especially if carbon sequestration techniques work. Global carbon emissions have doubled since 1970, and are now a quarter higher than in 2000, mainly due to the economic expansion of China and India.[5] Thus unless there is some dramatic world event progress to a low-carbon society will be very slow, if at all.

What about energy efficiency?

A common element of most visions of a low-carbon society is that we will, through energy efficiency measures, use less energy; this in turn will make such a society easier to achieve.[6] This belief that promoting energy efficiency will lead to lower national (or global) consumption is challenged by the 'rebound effect'. This argues (from economic theory and evidence) that lowering the cost of energy services, as energy efficiency does, leads to an increase in energy demand; this can (on the macro-scale) outweigh much, if not all, of the original savings (Herring and Sorrell, 2009; Herring, 2011). This is, naturally, a much-disputed topic between economists and environmentalists (Jenkins et al., 2011; Potts and Burns, 2011). What can apply to energy consumption, can also apply to all material consumption. Measures to cut carbon consumption, such as through behavioural change, can lead to increased consumption elsewhere. Thus (money) savings on home heating through turning down the thermostat, could lead to an increase in car travel, as the money saved on heating is re-spent. A recent modelling study from RESOLVE, the Surrey University research group, found that, on average, one-third of the carbon savings are re-spent (Druckman et al., 2011). They found there was a wide range of possible rebounds, from zero to one hundred per cent, according to what the money was re-spent on. It was possible for all the savings to be re-spent on carbon-intensive activities such as travel, (causing 'backfire') but also none if the money is invested in low-carbon technologies, such as solar PV panels. Thus it is essential that any money saved through lower consumption be redirected into low-carbon projects.

Encouragingly the UK government has started schemes to integrate efficiency and renewables. An innovative approach is PAYS (Pay As You Save) which gives households the opportunity to invest in energy efficiency (such as solid wall insulation) and micro-generation technologies (such as solar panels) in their homes with no upfront cost. Householders make repayments spread over a long enough period so that repayments are lower than their predicted energy bill savings; meaning financial and carbon savings are made from the beginning. The two-year pilot scheme was launched in 2009 for around 500 homes. A variety of measures have been installed including external wall insulation, solar photovoltaic panels, cavity wall insulation and boiler upgrades (EST, 2010). Thus the money savings from insulation are being used to subsidise renewable energy (rather than spent on greater consumption), and hence achieves lower emissions. This pilot scheme is now closed but is being replaced by the Green Deal programme (due to start in late 2012), which is discussed in Chapter 3.

Slow transport

We have a strong desire to travel, and this can be very carbon-intensive if we are in a hurry to reach our destination. Once travel was slow, by foot, horse or sail, and the journey was an integral part of the travel experience. Now we want to get to our destination as fast as possible, and the journey (often by cramped aircraft) is not an enjoyable experience but something to be endured. Aircraft are now the fastest growing source of carbon emissions.[7] It is hard to see how these emissions can be reduced given the lack of (cheap) alternative liquid fuels and our strong demand for quick (and cheap) travel to distant locations – typified by UK low-cost airlines offering long weekend breaks to Spain or Cyprus.

If the demand for travel cannot be curbed, then perhaps its nature can be changes, towards the emerging concept of 'slow travel'. This is derived from the slow food and the slow cities movement, that originated in Italy, and has been defined as 'an alternative to air and car travel, where people travel to destinations more slowly overland, stay longer and travel less' (Dickson and Lumsden, 2010). This concept fits into the goal of achieving transport sustainability, as expressed in Chapter 5, by reducing both trip length and in a shifting to less carbon-intensive modes of transport.

The feasibility of such a shift is explored by Warren and Ieromonachou (2011), who calculates the carbon emissions of a return trip from London to Cyprus. They contrast flying with going by rail (to a Mediterranean port, Marseilles, or Naples) and then by ship (see Table 16.1).

The return trip by air takes only ten hours, compared to over five days by rail and ship. The carbon intensity (kg CO_2 per passenger-km) of rail is very low, about a fifth that of air, while the intensity for ship travel is higher than air. So the greater the proportion of the journey done by rail, the lower the total emissions. Travel by air results in lower emissions than taking the boat from Marseilles, but more than taking it from Naples. Naturally, and the major advantage for passengers, is the speed of the air journey; it saves five days compared to the rail and boat journey.

However the train and the boat provide the passenger with all their energy needs, thus saving on the carbon emissions of a hotel room, estimated at about 30 kg CO_2 per day.[8] Thus, the carbon savings of travelling by air, compared to by ship from Marseilles, are swallowed up by the five days spent at the holiday resort (plus the additional carbon expenditure on local transport). Therefore the key question in evaluating travel modes

Table 16.1 London-Cyprus return by fast and slow travel

	Distance (kms)	Total CO$_2$ (Kgs)	Time (hours)	kgs CO$_2$ per passenger-km	Speed per hour	Km/hour
Air	6551	711	10	0.11	74	678
Rail to Marseilles, Ship to Cyprus						
Rail	1276	28	7	0.02	4.0	184
Ship	2937	409	60	0.14	6.8	49
Total	8426	873	134	0.10	6.5	63
Rail to Naples, Ship to Cyprus						
Rail	2103	57	14	0.03	3.9	146
Ship	2025	282	50	0.14	5.6	41
Total	8256	678	129	0.08	5.3	64

Source: Based on data from Tables 2 & 3, in Warren and Ieromonachou, 2011.

is what is done with the time saved. Generally it is used on carbon-intensive activities: car hire, day cruises, and staying in (air-conditioned) hotels. Few people fly to Cyprus and then go camping by bike!

The message from this data is that long-distance train travel has by far the lowest carbon-intensity, and the greater the proportion of the journey that can be done by it, the better. This 'slow travel' will, however, require a cultural shift away from placing primacy on speed of journey towards quality of journey, as is happening with the shift from 'fast food' to 'slow food' (Andrew, 2008).

Making the transition

The world has been talking about reducing carbon emissions for 20 years, but has little to show for it. Carbon (or more strictly speaking greenhouse gas) emissions have declined slightly in the UK and Europe, and has, mainly thanks to industrial restructuring, achieved the Europe's eight per cent reduction target for 2010. But the scale of reductions is nowhere near fast enough if we are to achieve a low-carbon society by 2050. The technologies are available, but the political will (and public support) to impose them is lacking. Immediate problems, like the recession, terrorism and the war in Afghanistan, dominate politicians' attention, and of course there are a whole host of worthy issues that fight for public attention, such as world poverty, care for the aged, or democracy in the Middle East. The threats posed by climate change are, to most people, far off and uncertain and while most people are sympathetic they are unwilling to do anything that is inconvenient or costly. With a recent Foresight report claiming that 'the UK ... is not expected to experience significant adverse direct climate change effects over the next two decades' little will be achieved unless the government places the country on a 'war footing' that allows for 'bold actions' (Foresight, 2011: 15, 113).

Small groups are, however, trying to make a start in creating low-carbon and sustainable communities, perhaps best typified by the Transition Town movement (Hopkins, 2008; Peters et al., 2010). Their efforts are popular as long as they bring consumer choice: the farmers market as well as the supermarket, solar panel in addition to the gas boiler, and the electric car alongside the petrol one. However, when their efforts intrude on other people's 'property' rights, such as with the erection of wind turbines or curbs on car use, there can be strong opposition.[9]

Strong government action, backed up by legislation, can overcome hostile public attitudes, as happened with the smoking ban and is happening with recycling. Spurred on by European directives, local councils have implemented, in the last decade, sweeping changes to rubbish collection and disposal, despite much adverse political and media comment on 'bin taxes'.[10] Waste going to landfill has nearly halved since 2000, with household recycling rates now at 40 per cent (compared to about 10% in 2000) and business rates at more than 50 per cent; the Environment Secretary, Caroline Spelman, talks about the UK having a 'zero-waste economy' (Defra, 2011). This regulatory action on waste disposal contrasts with the largely voluntary (in)action on carbon emissions. The results show clearly the impact of government directives, which can be enforced and implemented through local councils. Left to voluntary groups, nothing much can be achieved, as they are unable to overcome populist media hostility and vested interests. They can, however, create political and social support for radical, even unpopular, measures, which can eventual be enacted in legislation.

Future visions

Forty years is a long time in politics. Forty years ago, the National Union of Mineworkers and the UK Atomic Energy Authority were the most powerful energy institutions. Both were dedicated to furthering their interests. Both were laid low by a strong-willed prime minister, Margaret Thatcher, who wanted to reshape the energy landscape of Britain. This involved great political and social cost, and a complete restructuring of our energy industries; gas replaced coal, and private (and largely foreign) companies replaced nationalised industries. In the next 40 years, such a transformation could once again be seen, if there was another radical prime minister determined to see a low-carbon Britain. The power of our oil and gas companies could be curtailed, and their energies and money instead diverted into renewable energy and low-carbon technologies. There could be community co-ops providing local power and heat, solar PV on most rooftops, and farms producing bio-fuels. A super-grid could connect the geothermal power of Iceland to the solar power of the Sahara. Ultra-high speed trains could link all European cities, and airports would be largely redundant. Cities would be car-free; bikes everywhere and sailing ships very desirable holidays. The green utopia, or a Samsø version of it, would have arrived.[11]

Alternatively, we could live in a failed state like Iraq or Libya, blessed with (renewable) energy riches but unable to fully utilise them due to endemic social conflict. With no grid, communities would have to be self-sufficient, and travel would be difficult and dangerous. Life would be dominated by security of supply issues, in protecting one's own energy and food resources. Liquid fuels could be very expensive and highly prized, and long-distance travel would only be for the wealthy. Within communities solar and wind power would enable a basic standard of living, subject to periodic rationing. Life would be simpler, but not as we now wish it to be.

Whatever the future, a world of energy abundance or shortages, renewable energy will have a large part to play. This is because they are freely available everywhere, and unlike fossil fuels not concentrated in certain locations. As oil and gas run out, we can expect increasing regional conflicts over access to them (Klare, 2011). Rather than be drawn into 'oil wars', perhaps it would be better to turn our attention to our renewable energy riches. In the 1930s our grandfather's dream was of cheap electricity from hydro; the massive investment was beyond the capability of the private sector. Governments provided the will and the money, and now we enjoy the benefits. Similarly, renewables now require massive investments, and hopefully in 2050 our grandchildren will congratulate us on our sound investment.

Notes

1. New Testament, Mark 10: 25.
2. As interpreted by Margaret Thatcher in her quote 'No-one would remember the Good Samaritan if he'd only had good intentions; he had money as well.' TV Interview, 6 January 1980.

3. WWF (2011) report estimates them for the UK as £450 billion by 2025 – or about £30 billion a year over next 15 years.
4. Average domestic energy prices rose 126per cent between 2000 and 2009, with gas prices up 160per cent and electricity prices by 90per cent. Calculated from 'DUKES 2010: long-term trends', Tables 1.15: Consumption and 1.1.6: Expenditure, www.decc.gov.uk/en/content/cms/statistics/publications/dukes/dukes.aspx.
5. The IEA estimates that global carbon emissions rose 5.9 per cent in 2010, with three-quarters of the growth coming from emerging economies such from India and China, www.reuters.com/article/2011/05/30/us-iea-co-idUSTRE74T4K220110530.
6. For instance, WWF (2011, p. 5) has global energy demand 15 per cent lower in 2050 than it was in 2050, through 'ambitious energy-saving measures allow people to do more with less.'
7. Aviation fuel use in Europe grew by over 70 per cent between 1993 and 2008 (EuroStats 2011).
8. See www.carbonfund.org/site/pages/carbon_calculators/category/Assumptions.
9. The proposal for two wind turbines outside Totnes – the hometown of the Transition movement – has run into opposition from local objectors.
10. This was an issue at the 2010 general election, the Conservatives being strongly opposed to any charging schemes for waste collection or so called 'bin taxes'.
11. Samsø is Denmark's Renewable Energy island and is totally powered by wind energy, see the 10-year report – www.energiakademiet.dk/images/imageupload/file/UK/RE-island/10year_energyrepport_UK_SUMMARY.pdf.

References

Andrews, G. (2008) *The Slow Food Story: Politics and Pleasure*, London: Pluto Press.

Cambridge Carbon Footprint, www.cambridgecarbonfootprint.org.

Church, C. (2011) 'Creating Low Carbon Communities: A Report on Key Issues Arising from the Communities and Climate Action Conference', January 2011, www.lowcarboncommunities.net/2011-conference-report/.

Czisch, G. (2010) 'Scenarios for a Future Electricity Supply: Cost-Optimised Variations on Supplying Europe and Its Neighbours with Electricity from Renewable Energies', London: Institution of Engineering and Technology.

Defra (2011) 'Plans for a Zero Waste Economy Launched', *Defra News*, 14 June 2011, www.defra.gov.uk/news/2011/06/14/waste-review-published/.

Dickinson, J. and L. Lumsdon (2010) *Slow Travel and Tourism*, London: Earthscan.

EST (2010) *Home Energy Pay As You Save Pilot*. London: Energy Saving Trust, www.energysavingtrust.org.uk/Home-improvements-and-products/Pay-As-You-Save-Pilots, accessed 15 March 2011.

Eurostats (2011) 'Energy Statistics – Supply, Transformation, Consumption', epp.eurostat.ec.europa.eu/portal/page/portal/statistics/search_database.

Druckman, A., M. Chitnis, S. Sorrell and T. Jackson (2010) 'An Investigation into the Rebound and Backfire Effects from Abatement Actions by UK Households', RESOLVE Working Paper 05–10.

Foresight (2011) 'International Dimensions of Climate Change', The Government Office for Science, London.

Harper, P. (2007) 'Sustainable Lifestyles of the Future', in D. Elliott (ed.) *Sustainable Energy: Opportunities and Limitations*, Houndsmills, Basingstoke: Palgrave Macmillan, pp. 236–60.

Herring, H. (2011) 'Dealing with Rebound Effects', In *eceee 2011 Summer Study Proceedings*, Volume 1, Stockholm, Sweden.

Herring, H. and S. Sorrell (eds) 2008 *Energy Efficiency and Sustainable Consumption: The Rebound Effect*, Houndsmills, Basingstoke: Palgrave Macmillan.

Hopkins, R. (2008). *The Transition Handbook: From Oil Dependence to Local Resilience*, Totnes: Green Books.

Jenkins, J., T. Nordhaus and M. Schellenberger (2011) 'Energy Demand Backfire and Rebound as Emergent Phenomena: A Review of the Literature', Oakland, CA: Breakthrough Institute.

Klare, M. (2011) 'Energy: The New Thirty Years' War', *Guardian Comment Network*, 29 June 2011, www.guardian.co.uk/commentisfree/2011/jun/29/energy-war-global-conflict.

Marshall, G. (2007) 'Can this Really Save the Planet?' *The Guardian*, 13 September 2007, www.guardian.co.uk/environment/2007/sep/13/ethicalliving. climatechange.

Michaelis, L. (2006) 'Ethics of Consumption', in T. Jackson (ed.) *Sustainable Consumption*, London: Earthscan, pp. 328–45.

Peters, M., S. Fudge and T. Jackson (2010) *Low Carbon Communities: Imaginative Approaches to Combating Climate Change Locally*, London: Edward Elgar.

Potts, M. and C. Burns (2011) 'The "Rebound Effect": A Perennial Controversy Rises Again', *Solutions*, Spring 2011, blog.rmi.org/ReboundEffectPerennialCon troversyRisesAgainSJBlog.

Shields, R. (2010) 'Green Fatigue Hits Campaign to Reduce Carbon Footprint', *The Independent*, 10 October 2010, www.independent.co.uk/environment/climate-change/green-fatigue-hits-campaign-to-reduce-carbon-footprint-2102585.html.

Skea, J., P. Ekins and M. Winskel (eds) (2010) *Energy 2050 – Making the Transition to a Secure Low Carbon Energy System*, London: Earthscan.

Warren, J. and P. Ieromonachou (2011) 'Slow Cyprus – Alternative Tourist Routes', paper to 3rd International Conference on Special Interest Tourism & Destination Management, Kathmandu, Nepal, 27–30 April 2011.

WWF (2011) 'The Energy Report – 100% Renewable Energy by 2050', WWF & Ecofys, wwf.org.uk/energyreport.

CPSIA information can be obtained at www.ICGtesting.com
Printed in the USA
LVOW03*0019160115

422994LV00009BB/71/P

DATE DUE